The EU Emission Trading Scheme

European University Studies

Europäische Hochschulschriften

Publications Universitaires Européennes

Series V

Economics and Management

Reihe V Série V

Volks- und Betriebswirtschaft

Sciences économiques, gestion d'entreprise

Vol./Bd. 3363

PETER LANG

Frankfurt am Main · Berlin · Bern · Bruxelles · New York · Oxford · Wien

Stefan Veith

The EU Emission Trading Scheme

Aspects of Statehood, Regulation and Accounting

PETER LANG
Internationaler Verlag der Wissenschaften

Bibliographic Information published by the Deutsche Nationalbibliothek
The Deutsche Nationalbibliothek lists this publication in the Deutsche Nationalbibliografie; detailed bibliographic data is available in the internet at http://dnb.d-nb.de.

Zugl.: Bremen, Univ., Diss., 2010

D 46
ISSN 0531-7339
ISBN 978-3-631-60315-4

www.peterlang.de

Acknowledgements

This study is based on my doctoral thesis, which was accepted by the Faculty of Business Studies and Economics at the University of Bremen in February 2010. My thanks go to a number of individuals.

In the first place, I am deeply grateful to my supervisor Prof. Dr. Dr. h.c. Jochen Zimmermann, University of Bremen, for his helpful advice, perpetual encouragement and individual guidance. Furthermore, I am extending my sincere gratitude to Prof. Dr. Michael Hülsmann, Jacobs University Bremen, for being my second referee. Without hesitation he agreed to review my thesis under difficult circumstances. Prof. Dr. Martin G. Möhrle, University of Bremen, and Prof. Dr. Jörg Prokop, University of Oldenburg, as well as my colleagues Tim-Frederik Oehr and Johannes Schymczyk participated in my dissertation committee. Many thanks for this.

My research largely benefited from the collaboration with Dr. Jörg R. Werner, University of Bremen. His comprehensive review of my writings and his challenging ideas together with lengthy but yet fruitful discussions gave rise to major improvements of this study. My warm thanks also go to the diehard optimist Stephan Abée, with whom I shared my office during our dissertation projects. I am also grateful to Luka Bichinashvili and Jonas Roggenbock. They supported me in manually collecting the vast amounts of financial and emission data which were necessary for my empirical analyses.

Finally, I owe many thanks to my parents Erich and Ursula Veith as well as to my partner Lena Borgmann, who encouraged and supported me throughout my entire dissertation project.

The financial support of the *Bereichsforschungskommission Sozial-/Geisteswissenschaften* at the University of Bremen and of the *Wiwib e.V. – Wirtschaft-Wissenschaft-Bremen (Verein zur Förderung des Fachbereichs Wirtschaftswissenschaft der Universität Bremen und dessen Zusammenarbeit mit der Bremer Wirtschaft)* is gratefully acknowledged.

Contents

List of Figures

List of Tables

List of Abbreviations and Symbols

AAU	Assigned Amount Unit (Kyoto)
ACER	Agency for the Cooperation of Energy Regulators (EU)
AEM	Azienda Elettrica Municipale (Italy)
ARC	Accounting Regulation Committee (EU)
Art.	Article
ASM Brescia	Azienda dei Servizi Municipalizzati di Brescia (Italy)
BGBl	Bundesgesetzblatt – Federal Law Gazette (Germany)
BMU	Bundesministerium für Umwelt, Naturschutz und Reaktorsicherheit – Federal Ministry for the Environment (Germany)
BVE_{it}	book value of equity of firm *i* in period *t*
${}^{adj}BVE_{it}^{q}$	book value of equity, adjusted for the ETS balance sheet item (either allowances asset or emission liability), scaled by the market value at fiscal quarter beginning, of firm *i* in quarter *t*
CAPM	capital asset pricing model
CDM	Clean Development Mechanism (Kyoto)
CCGT	combined cycle gas turbine
CEER	European Regulators' Group for Electricity and Gas (EU)
CER	Certified Emission Reduction (Kyoto)
CEGB	Central Electricity Generating Board (England and Wales)
CEPI	Confederation of European Paper Industries
CHP	combined heat and power
CITL	Community Independent Transaction Log (EU)
CO_2	carbon dioxide
COST	IFRIC 3 cost accounting treatment for emission rights
CR	concentration ratio
$DEMITT_i$	dummy variable equalling 1 if the share of fossil

	power generation for firm *i* is above sample median, and 0 otherwise
$DENDOW_{it}$	dummy variable equalling 1 if firm *i* was allocated more rights than it emitted during an annual cycle, and 0 otherwise
$DIFIRC^q_{it}$	dummy variable equalling 1 for IFRIC 3 adopters, and 0 otherwise
$DMKTCHG_i$	dummy variable equalling 1 if the concentration ratio in the country of incorporation of firm *i* fell more than the sample median, and 0 otherwise
DMO	Debt Management Office (UK)
$DNEG^q_{it}$	dummy variable equalling 1 for a negative ETS earnings impact during fiscal quarter *t* (i. e. $ETS\_EARN^q_{it} < 0$), and 0 otherwise
DONG	Dansk Olie og Naturgas (Denmark)
$DPRGRW_i$	dummy variable equalling 1 if the mean annual growth in electricity production in the country of incorporation of firm *i* is above sample median, and 0 otherwise
DR^q_{et}	dummy variable equalling 1 for negative returns on emission allowances during fiscal quarter *t* (i. e. $R^{\delta q}_{et} < 0$), and 0 otherwise
DSO	distribution system operators
$DTARIFF_i$	dummy variable equalling 1 if the independent regulatory authority in the country of incorporation of firm *i* is qualified to determine end user prices, and 0 otherwise
$EARN_{it}$	earnings of firm *i* in period *t*
$EARN^q_{it}$	earnings before extraordinary items of firm *i* in quarter *t*, scaled by the market value at fiscal quarter beginning
$^{adj}EARN^q_{it}$	adjusted earnings before extraordinary items (i. e. reported income plus/minus ETS earnings impact according to one of the four alternative treatments), scaled by the market value at fiscal quarter begin-

	ning, of firm *i* in quarter *t* (*i. e.* $EARN_{it}$ - ETS_EARN_{it})
$EARNCH_{it}$	change in earnings with respect to the same period of a year earlier of firm *i* in period *t*
$EARNCH^q_{it}$	change in quarterly earnings before extraordinary items with respect to the same quarter of a year earlier, scaled by the market value at fiscal quarter beginning, of firm *i* in period *t*
ECX	European Climate Exchange (UK)
ÉDF	Électricité de France (France)
EEX	European Energy Exchange (Germany)
EFRAG	European Financial Reporting Advisory Group (EU)
EFTA	European Free Trade Association
EHVV	Emissionshandels-Versteigerungsverordnung – Emission Trade Auction Regulation (Germany)
EnBW	Energie Baden-Württemberg (Germany)
ENEL	Ente Nazionale per l'Energia Elettrica (Italy)
EnWiG	Gesetz zur Förderung der Energiewirtschaft – Energy Economics Act (Germany)
ERC	Emission Reduction Credit (Kyoto)
ERGEG	European Regulators' Group for Electricity and Gas (EU)
ERU	Emission Reduction Unit (Kyoto)
ETS	Emission trading scheme
$ETS\_BSI^q_{it}$	ETS-specific balance sheet item (either allowances asset or emission liability) according to one of the four alternative treatments of firm *i* in period *t*, scaled by the market value at fiscal quarter beginning
$ETS\_EARN^q_{it}$	quarterly net ETS earnings impact according to one of the four alternative treatments, scaled by the market value at fiscal quarter beginning, of firm *i* in period *t*
EUA	European Union Allowance
EXAA	Energy Exchange Austria (Austria)

FASB	Financial Accounting Standards Board (US)
FFV	proposed full fair value accounting treatment for emission rights
FW	IASB Framework
GDF	Gaz de France (France)
HEW	Hamburgische Electricitäts-Werke (Germany)
i	firm index
IAS	International Accounting Standard
IASB	International Accounting Standards Board
IFRIC	International Financial Reporting Interpretations Committee
IFRS	International Financial Reporting Standard
IPCC	Intergovernmental Panel on Climate Change (UN)
JI	Joint Implementation (Kyoto)
LEV_{it}^{q}	leverage – total debt divided by total assets for firm i in quarter t
MTB_{it}^{q}	market-to-book ratio – market value at the end of quarter t divided by the book value of equity for firm i
NAP	National Allocation Plan (EU)
NBP	Nitrogen Oxides Budget Trading Program (US)
NEPI	new environmental policy instrument
NET	USofA net accounting treatment for emission rights
NO_x	nitrous oxide
NRA	National Regulatory Authority (EU)
Ofgem	Office of Gas and Electricity Markets (UK)
OJ	Official Journal of the European Union
OTC	Ozone Transport Commission (US)
P_{it}	price of the common stock of firm i in period t
P_{it}^{q}	price of the common stock ending one month after fiscal quarter end of firm i in period t, scaled by the price at fiscal quarter beginning
$R_{et}^{\delta q}$	discrete return on emission allowances during fiscal quarter t

${}^{X}R_{et}^{ld}$	logged daily return on emission allowances in period t
${}^{X}R_{ft}^{ld}$	logged daily return on local currency towards USD in period t
${}^{X}R_{gt}^{ld}$	logged daily return on Zeebrugge natural gas in period t
R_{it}	common stock return of firm i in period t
R_{it}^{δ}	discrete common stock return of firm i in period t
$R_{it}^{\delta q}$	discrete buy-and-hold common stock return for a three-month period ending one month after fiscal quarter end of firm i in period t
R_{it}^{ld}	logged daily return on the common stock of firm i in period t
R_{mt}	return of the market portfolio in period t
R_{mt}^{ld}	logged daily return on the market portfolio in period t
${}^{X}R_{ot}^{ld}$	logged daily return on crude oil in period t
RECLAIM	Regional Clean Air Incentives Market (US)
REV	IFRIC 3 revaluation accounting treatment for emission rights
RGBl	Reichsgesetzblatt – Imperial Law Gazette (Germany)
RMU	Removal Unit (Kyoto)
RSI	residual supply index
SO_2	sulphur dioxide
t	time index
TEHG	Gesetz über den Handel mit Berechtigungen zur Emission von Treibhausgasen – Greenhouse Gas Emission Trading Act (Germany)
TSO	transmission system operators
UNFCCC	United Nations Framework Convention on Climate Change
US GAAP	United States Generally Accepted Accounting Principles
USC	United States Code

USofA	Uniform System of Accounts (US)
VEAG	Vereinigte Energiewerke AG (Germany)
VEBA	Vereinigte Elektrizitäts- und Bergwerks AG (Germany)
VIAG	Vereinigte Industrieunternehmungen AG (Germany)
VWE	Vereinigte Elektrizitätswerke Westfalen (Germany)
X_{jt}	vector of additional explanatory variables observable in period t
ZuG	Gesetz über den nationalen Zuteilungsplan für Treibhausgas-Emissionsberechtigungen – Allocation Act (Germany)

1 Introduction

1.1 Motivation for the Research

With the introduction of the emission trading scheme (ETS) in 2005, European emission regulation took a major step towards market-based solutions. The new scheme's mechanisms are unlike the ones of traditional instruments like eco taxes, and regulatory bodies took some time to develop adequate rules, as two anecdotes highlight. At the inception of the programme the regulators (i. e. EU Member States and the EU Commission) had issued too generous a number of allowed emissions. The price for emission rights duly dropped to only some cents per tonne in early 2007. This issue of oversupply was fixed by 2008 when the legal framework was eventually changed. Another case of regulatory failure is the rise and fall of international accounting norms for emission trading schemes. After a first draft had been presented in 2003-05, its final version was adopted in 2004 by the competent standard setting organisation. By 2005-05, it became apparent that this rule set would not be endorsed to EU law, and it was finally withdrawn. Since then, the accounting community has been debating on how to appropriately map the economic incentives of an ETS in a financial report.

Both anecdotes illustrate that the mechanisms of an emission trade are not as straightforward as the policymakers had expected. The main reason is that a trading scheme is a policy instrument rooted in neoclassical economics that by design relies on the market. Emission trade is not only set out to mitigate pollution, but also to achieve an emission reduction that controls for the diverse polluters' cost structures. This central feature represents the scheme's strength and weakness at the same time: different firm sizes, input good mixes and demand patterns result in specific emission levels that are balanced by means of the price mechanism. As market prices are freely determined by its participants and are subject to constant fluctuations, the resulting net effect on the regulated firms cannot be easily predicted ex ante.

Blaming the regulatory authorities for an implementation failure ignores difficulties in the economic theory. Textbook examples for an ETS describe the pricing patterns as well as the regulator's tasks, given certain strict assumptions hold. At closer inspection, some of these assumptions are not in line with realities – more specifically, the neoclassical theory ignores institutions. Energy market arrangements that have traditionally been relying on different steering modes like state-owned or self-organised entities cannot be mapped in the model. Consequently, the institutional arrangements of two thirds of the firms currently subject to the scheme remain disregarded. Further, if there are no institutions, the firms do not need to report on their accomplishments in the energy or emission markets. There is no demand for accounting information regarding the market effects on the firm under scrutiny.

In sum, understanding the economic consequences of emission trading schemes is equally important from a policymaking as well as from an economic perspective. Further insights into its mechanisms are needed since this environmental instrument is already or about to be implemented under diverse conditions. Various schemes are in force in some regions of the US, an intergovernmental trade is prescribed in the Kyoto Protocol, and other countries like Australia are planning national schemes. The existing EU regulation will be further extended from 2012 onwards to cover the aviation and some branches of the chemical industries, and a significant amount of the emission budget will be auctioned off.

Figure 1: Research Questions

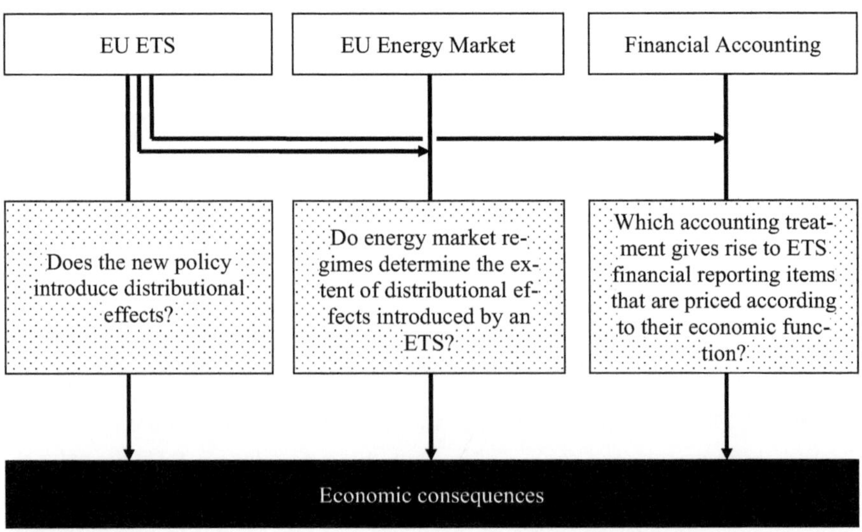

This study aims at closing gaps in research concerning underlying institutional arrangements and distinct economic consequences of the EU ETS. In doing so, it gives an overview of the available environmental policy instruments and the distinct characteristics of an emission trade. Incentive structures at business level are largely determined by means of accounting numbers – hence this analysis will additionally provide a review on alternative management and financial accounting treatments. As sketched in Figure 1, the study will furnish empirical evidence regarding three unresolved, but intertwined questions: Firstly, it is asked whether the EU ETS introduces distributional effects between the regulated energy firms and other market participants. It will be shown whether the scheme gives rise to an additional burden for the sector or whether the producers are able to compensate the costs. The method chosen is novel since it does not analyse demand and supply functions, but the investor expectations towards the ETS's net effect. Secondly, it is addressed whether energy market regimes determine the extent of distributional effects introduced by an ETS. This is a direct test for the impact of institutional arrangements on the uniform European regulation. Thirdly, it is analysed which financial accounting treatment gives rise to ETS financial reporting items that are priced according to their economic function. Since there is presently no specific international accounting standard for this issue in force, several alternatives may be employed or were proposed. Answering this question again takes a capital market perspective and shows which of the treatments is able to furnish useful information for investors.

The sketched research questions will be addressed by means of empirical analysis. Investigating into the investors' predictions towards the resulting economic effects of a regulation, changes in capital market expectations are used as gauge for changes in a firm's wealth. Consequently, the ETS is termed to lead to rent redistributions if the empirically found properties are not in line with the intended outcome. This is the case as soon as increments on the market for emissions lead to the expectation of increasing future profits. The results of this study suggest that the goal of making pollution more costly was not achieved during the first three years of emission trading. Instead, energy firms were found to over-compensate for the regulatory costs. Additionally, the degree to which these rents are shifted to customers or non-energy firms seems to be determined by the underlying institutional regime of energy markets. According to the empirical evidence of this analysis, utilities in market-based arrangements generate the largest windfall profits – indicating a lacking maturity of the recently created regimes.

Contrasting the information content of four financial accounting treatments also relies on examining the capital market expectations. Here, e. g. accounting revenues (expenses) concerning the ETS should lead to positive (negative) pricing effects. Results for the first three trading years suggest that those treatments bearing technical deficiencies, so called mismatches, are not suitable to convey relevant information for equity valuation purposes. Two options were identified to overcome this issue: either by only reporting on the regulatory net effect or by mitigating the mismatches using adjusted reporting methods.

1.2 Structure of the Analysis

The remainder of this study is organised as follows. In Chapter 2, the new regimes of European environmental and energy market regimes are depicted. These policies are classified by means of a novel approach which takes its start from the involved actors that can provide for welfare: the state, the market or the community. In this respect, the regulatory costs of pollution control are determined by the state under all classical state-based environmental instruments, whereas the market settles the abatement costs only under an ETS. An overview is given how the EU implemented the economic concept of emission trading to the present legal framework, and which national interests lead to an unbalanced setup. This also includes discussing the question why the Member States embarked on this new regulatory approach, and institutional set of causes is identified as driving this political process. Further, due to its dominance within the EU ETS, the European energy market regulation is illustrated. Here, the institutionalist approach is extended to the different national regimes of generating and supplying electricity. It is shown that the currently ongoing transformation towards a market-based design was triggered as early as 1989. Consequently, analysing the political incentives of introducing a market-based environmental pollution control cannot be regarded separately from analogous steps in energy market reconfigurations. The chapter is summarised by the first two empirical research questions asking for the direction of economic effects and the determinants of market regimes. Both of them will be answered in Chapter 4.

Chapter 3 picks up the issue of how to account for emission trading schemes. If this regulation is set out to develop economic incentives, they necessarily need to be reflected in the financials at business level. The two different reporting systems, management accounting and financial accounting, are discussed with respect to their common stewardship and decision making functions. Since management reporting is not determined by standards or laws, the accountant is free to choose between imputed and accrual costs. Whereas imputing costs always furnishes relevant information for both functions of accounting, the accrual approach fails in case of large free emission budgets. Financial reporting, for its part, is strictly regulated for listed firms. In the special case of emission trading schemes, several distinct treatments can be equally applied until a final standard will be eventually adopted by 2011. The major four treatments are juxtaposed using a case study, allowing for the illustration of distorting effects on interim accounting income. The normative analysis results that the stewardship function of accounting is best fulfilled when netting the ETS effect, and that the decision making function is best reflected when introducing a modified measurement approach (full fair value accounting). This conclusion is the starting point for the third empirical research question asking for the pricing of ETS-related accounting items.

Chapter 4 develops the models and derives the samples employed in the empirical analyses to address the open research questions. In doing so, a multifactor market model is estimated to gauge the distributive effects of the EU scheme in the power sector. It furnishes evidence for windfall profits being predicted by the capital market – a result being robust controlling for commodity prices and regulatory conditions. Exceptions are utilities that largely rely on fossil fuels; they seem to be exposed towards additional regulatory costs. Finally, the model is estimated controlling for the dominant energy market regime. It is found that state-based arrangements mitigate the extent of windfall profits, whereas the market-based design yields the largest over-compensation for electricity producers. Consistently, different capital market exposures with respect to the energy market regimes can also be found prior to the introduction of the EU ETS. This chapter concludes with the assessment of accounting items arising from four different treatments for trading schemes using a value relevance model. Balance sheet items are consistently priced in all of these alternatives, but the respective earnings give rise to differences. This effect is attributable to the treatments withdrawn due to their distorting measurement effects. These results remain robust when controlling for risk factors and emission price changes.

Chapter 5 summarises the results of this study.

2 Transformation of Environmental and Energy Policies in the European Union

2.1 Towards a Governance Framework for Policy Analysis

Most existing emission trading schemes are based on a budget, which caps the pollution of a substance to a predefined volume per period. This budget is subdivided into a number of tradable certificates, of which each carries the same allowance.[1] At the end of the period, the regulated entities have to surrender the amount of certificates that corresponds to the physical emissions. Depending on the regulatory details, the regulated entity can receive the certificates by an agency in advance (by means of a free or a costly allocation) or it can purchase necessary or sell supernumerous units on designated exchanges during the period. This type of an emission trading scheme is usually called a cap-and-trade mechanism.

Such an ETS deviates from a traditional regulatory policy, which relies on precise standards, or direct financial (dis)incentives. As a new policy mode it is at the same time preferred and criticised for its unique features. Supporters underline its economic efficiency (Stavins, 1998: 70-71), whereas critics regard it as intrusion of neoliberal ideas to environmental politics (Bailey, 2007: 530). The current debate on emission trading schemes and their comparison to other instruments is split into economic and political sciences schools of thought. The former strand of literature identifies an efficient, least cost approach – neglecting constraints arising from the instrument's institutional setting. The latter focuses on its political and cultural embeddedness – largely disregarding economic incentives.

The following policy analysis illustrates the regulatory context of emission trading schemes: environmental regulation. In doing so, it is necessary to describe the relevant actors both configuring and subject to the available regulatory alternatives. Distinguishing their respective scopes and incentives yields quality criteria for evaluation. This analysis will develop a taxonomy of policy

---

1 The tradable allotment of pollution is either called right, allowance, certificate or unit. Throughout this analysis, the named terms are used interchangeably.

regimes which helps to understand transitions in regulation over time and serves to show linkages to other fields. In doing so, it takes a business perspective and aims at considering economic incentives and institutional arrangements at the same time.

Most authors distinguish between command-and-control instruments and incentive-based instruments[2] (see only Kolstad, 1986: 263; Hahn and Stavins, 1992: 464). The underlying intuition is as follows: command-and-control regulation, like technology or performance standards, directly intervenes in production processes to achieve a desired volume of pollution (Goulder and Parry, 2008: 157-159). Incentive-based instruments, like taxation, negotiation, subsidies, or emission trading schemes have an indirect environmental effect setting economic incentives via pricing in- or decrements (Lewis, 1996: 820). This allows to predict the behaviour of regulated agents, and hence to define quality criteria. In providing for economic efficiency, the incentive-based approach is perceived to be largely superior to command-and-control instruments. Some research classifies environmental policy regimes by means of a market-based versus non-market based dichotomy (Bailey, 2007: 532). In this case, taxation and ETSs are differentiated from the other named instruments since their price increment comes into effect on the product market. Both classifications are marked by inconsistencies that cloud a clear view on the institutional setting of environmental policies. For example, it is questionable why taxation should be regarded as a market-based solution, whereas subsidies should not. Further, grouping both taxation and emission trading as incentive-based instruments is contradictory from an institutionalist perspective. Whereas a market scheme exhibits a considerable retreat of the state, levies are the basis to a strong tax state (Genschel, 2005: 55). The determination of the regulatory costs, and hence the factual pollution reduction, takes place without direct public interference under trading or negotiation schemes.

2 The political sciences literature prefers the notion of new environmental policy instruments (NEPIs) instead of incentive-based regulation (e. g. Wurzel et al., 2003: 51). Both concepts basically mean the same, but as opposed to economists, political scientists also include non-pecuniary instruments like eco labels (Jordan et al., 2003: 212-213). For the sake of consistency, the economic terminology and meaning is used throughout the remainder of this study.

Applying the Streeck and Schmitter (1985) classification of institutional arrangements to environmental policy regimes seems more promising to achieve a clear distinction of economic policies. It differentiates between three[3] modes of regulation: state, community and market (for a critique see Schuppert, 1990: 232-234). According to this approach, the state relies on a hierarchical control and exerts its power by means of the bureaucracy, and both other bases of social order involve certain degrees of self-regulation by private actors. The community is based on spontaneous solidarity among a network of agents with similar interests, and markets are characterised by profit maximisers and dispersed competition (Streeck and Schmitter, 1985: 120-121). Adopting this taxonomy to environmental policy gives the following, different classification of the available policy instruments.

In a statist approach, environmental policies are chosen as specifications regarding technology and performance standards, subsidies and taxes, as these are the instruments determined and enforced by the government or other public bodies. The communitarian setting implements environmental policies as negotiation instruments by polluters and victims. Even though compensation schemes can be stipulated by law, the actors involved reach agreements independently. Finally in a market approach, policies take the form of abatement measures which are bound to the trade in emission rights mechanism balancing firm-specific long or short positions. Although a public authority enacts and supervises provisions of the emission trade, its economic effect depends on the free settlement of demand and supply.

Table 1 summarises and gives an overview of the available environmental policies and the three classifications based on the concept of social welfare. Quite clearly it could be objected that all kinds of policy regimes can be traced back to state interventions and rely on public coercion. Of course, nearly no imaginable regulation embodies the archetypical characteristics in an ideal form, and the same criticism can be voiced over similar adoptions of the institutional arrangements approach, as e. g. proposed for accounting regulation in Puxty et al. (1987), Volmer et al. (2007), and Zimmermann et al. (2008). Indeed, it is possible that a technology standard is elaborated also in cooperation between the

---

[3] Even though Streeck and Schmitter (1985) define associations as a fourth social order, they are commonly regarded as a subset of the community (Volmer et al., 2007: 440).

government and an industrial association, whereas the market mechanism only develops within the strict rule set of the state. Consequently, this tripartite classification mainly asks for the general actor providing for welfare, and in the case of an ETS, this is clearly the market for allowances. In contrast to that, the welfare effect of taxation draws on the allocation and redistribution of the treasury, resulting in aligning eco taxes as genuine state function. The proposed taxonomy promises to be the most suitable framework for the remaining analysis of environmental policy regimes. It makes it possible to isolate the scope and incentives of the respective promoting actor and in the end to evaluate the emerging frameworks. In the case of an ETS, it serves to explain the causes promoting a transformation of statehood, i. e. why and by which means state-based approaches are reconfigured to market-based regimes.

Table 1: Taxonomy of Environmental Policy Regimes

Welfare perspective	Classification		
Economic efficiency	*Command-and-control*	*Incentive-based*	
	Technology standard Performance standard	Taxation Negotiation Subsidy Emission Trade	
Mechanisms of wealth transfer	*Non-market-based*	*Market-based*	
	Technology standard Performance standard Subsidy Negotiation	Taxation Emission Trade	
Provider of welfare	*State*	*Community*	*Market*
	Technology standard Performance standard Subsidy Taxation	Negotiation	Emission Trade

Environmental policy in many cases targets energy market – in case of the EU ETS, power generation accounts for roughly 65% of all CO_2 emissions. That is why the respective constitution of the energy markets should not be neglected. Energy markets are usually classified according to their degree of competition (e. g. Meeus et al., 2005: 30-32; Domanico, 2007: 5067-5068): A monopoly with either state-owned enterprises or a network of regional monopolists exhibits the lowest intensity of competition, followed by an oligopoly that can be found in early deregulated markets, and by a polypoly with a (perfectly) competitive market. This taxonomy is able to explain different incentives and conduct of market participants, and testable hypotheses can be deduced from it.

Table 2: Taxonomy of Energy Market Regimes

Welfare perspective	Classification		
Degree of competition	*Monopoly* State-owned enterprises Network of entities	*Oligopoly* (Early) deregulated market	*Competition* Competitive market
Provider of welfare	*State* State-owned enterprises	*Community* Network of entities	*Market* Competitive market (Early) deregulated market

As providers of welfare extend to society at large (Hollingsworth, 2000: 601) it is possible to adopt the above modes of governance to energy market regimes, too. Prior literature only implicitly relies on this classification, like Humphreys and Padgett (2006: 387, 389), who characterise the power markets of France as state-dominated and of the UK as market-dominated with a decreasing intensity of regulation. Eberlein (2000: 84) finds that the German corporatist mode of energy regulation relies on a network of private and state-owned entities. In this respect, the bases of social order – state, market, community – can already be found to a certain extent in this domain of regulation, as summarised in Table 2. This will now be developed in a systematic fashion.

In a statist approach, energy generation, and to some extent even price determination, is pooled in one state-owned enterprise. The communitarian setting is characterised by a number of corporations partitioning the market as well as the technological and economic (i. e. price determination) framework. To a certain degree, public actors (ex ante) intervene or (ex post) codify important arrangements. Finally in a market, several suppliers compete via the price mechanism. During the changeover period from the state or communitarian to the market mode (i. e. in early deregulated markets), a strong public enforcement agency provides for equal usage and access rights.

Applying the framework of social order to environmental and energy market regulation does not only allow for a novel taxonomy. It is additionally able to identify promoting actors within the respective policy approaches and to outline causes determining or explaining some of the regulatory outcome. Since the major regulatory actor is the state, it is asked which tasks in these policy fields can be conceptually and are factually assumed by the state or by other arrangements. In the post-war period, a strong and interventionist nation state could be found throughout Europe, which bundled the law, legitimacy, the provision of welfare and the territorial integrity. In the 1980s, this so called golden age of the nation state came to an end (Hobsbawm, 1995: 325; Leibfried and Zürn, 2005: 5). Since then, the conception of policymaking has transformed towards a stronger emphasis on market-based arrangements. In this sense, the state as a setter of strict rules disappears, or more specifically, the scope of responsibilities taken on by the state in order to provide for welfare shifted since then. Three types of public responsibilities – regulatory, operational and outcome responsibility – are now dispersed across the actors state, community and market (Zimmermann, 2008: 174-175). A state fulfilling the first category uses the legal instruments and powers in providing welfare. Operational responsibility additionally necessitates that a public authority is entrusted with implementing the rules. Outcome responsibility finally applies if the state acts as corrective when welfare is perceived to be distributed unequably.

A golden age environmental or energy market policy regime is carried out by the state if it assumes all three types of responsibility. In doing so, typically the state-based social order predominates in both fields of interest. If a transformation of statehood can be observed after the hypothesised turning point, e. g. market-based environmental or energy market regimes are likely to emerge. Here, the public sector is ideally only in charge of setting up a functioning emission right market. Operational responsibility is not necessarily needed, as private actors can assume this function. Because welfare is solely provided for by the market, no outcome responsibility is found with the state. However observing compliance with public interest, the state will not refrain from its regulatory responsibility (Volmer et al., 2007: 453).

Sections 2.2 and 2.3 will now take a closer look to the fields of environmental and energy market regulation, respectively. In the following analysis, the available regulatory modes are assessed with regard to their economic theory, resulting legal framework, and implications arising from institutional fits or divergences. This gives rise to evaluation criteria required when analysing the relative shortfalls and merits of the policy regimes and when addressing the changes in statehood in both policy fields throughout the EU Member States.

2.2 Environmental Regulation

2.2.1 Economic Theory of Pollution Control

2.2.1.1 General Setting

The economic justification of regulating pollution is to avoid market failure. A market failure is captured by the concept of so called external effects (or externalities) arising whenever an economic agent does not pay for the usage of a resource. He thus transfers an economic burden to unrelated parties that are not adequately compensated for the damage (Baumol and Oates, 1975: 18). This resource misallocation leads to an interference with the utility or production functions of the affected individuals or firms, respectively (Mishan, 1971: 2). In case of pecuniary externalities, a transmission is accomplished by means of the price mechanism, whereas technological externalities are not transmitted via the market (Viner, 1932: 40-41; Scitovsky, 1954: 144). A common example for the pecuniary type is the increased production of a certain good by one firm, ceteris paribus leading to a price decease to all other competitors in a neoclassical world. Adverse effects arising from pollution are representatives for a technological externality (Liebowitz and Margolis, 1994, 1994: 137).

Since the market participants are unable to solve this problem on their own, another institution is needed to internalise the aggregate external effect. Traditional welfare economics approaches assume only the state is able to transfer the economic burden back to the originator, but an efficient, least-cost regulation has to control for firm-specific cost structures in production and abatement. Evaluation criteria other than aggregate efficiency – like ecological effectiveness or administrative costs (arising from monitoring and enforcement) – are mostly disregarded in environmental economic analyses (Weersink et al., 1998: 312). Further, the economics of technology literature evaluates the policy maker's ability to promote innovation (e. g. Malueg, 1989; Jaffe and Stavins, 1995), but this criterion is usually subordinate in the welfare assessment of instruments.

2.2.1.2 Traditional Frameworks of State Intervention

To provide for welfare, the state can mandate technology or performance standards, grant subsidies or levy taxes. Technology standards either require a polluter to employ a certain end-of-pipe technique (like a filter) or the best available technology (like switching to cleaner fuels) in achieving a certain pollution reduction (Goulder et al., 1999: 339).[4] However, this solution suffers from the severe shortcoming that the regulator is faced lacking information. If it is the aim to achieve a least-cost emission reduction, the agency not only needs figures such as the entity-specific pollution, but also unobservable data like detailed parameters of the production process. Unless they are available, the costs in putting the new technology in operation vary over the population of firms of different ages and employing various processes (Cropper and Oates, 1992: 686). Another issue of technology standards is their inefficiency. A firm does not fully internalise the external effect because it is only coerced to invest in the mandated abatement technology without having to pay for the free usage of the sink (Goulder and Parry, 2008: 158). Consequently, this regulation directly mandates which technology is to be applied in the production process and does not leave any choice with the economic agents. The state enforces the introduction of new technologies that market participants otherwise view as too costly to adopt unilaterally for remaining competitive (Jaffe and Stavins, 1995: S-45). Instead of focusing on economic efficiency in terms of internalisation, the regulator aims at promoting innovation and the development of completely new technologies. To this end, asymmetrically distributed compliance costs among the regulated entities are accepted (Milliman and Prince, 1989: 248-249).

---

4 A specific type of technology standards is increasing liability rules (see only Garber and Hammitt, 1998); due to its similarities with respect to economic incentives and regulatory implications, it is not separately analysed in the following.

Next to technology standards, the second statist instrument – performance standards – is quite similar to emission trading schemes. Both instruments allow a certain maximum output of harmful substances, mostly expressed as a fraction per unit produced (Requate, 2005: 178). Introducing relative limits slightly relaxes the policy maker's bounded rationality and controls for demand fluctuations concerning the output good. However, regarding these volumes-based approaches one major difference emerges: the standard cannot be traded among the regulated entities. Consequently, economic shortcomings similar to technology standards apply in this case, too (Carlson et al., 2000: 1315-1316; Goulder and Parry, 2008: 158), and they are unable to provide for more welfare than other state-governed environmental instruments (Quirion, 2005: 351-352). Further, these performance standards induce aggregate compliance costs that do not (or only by chance) meet the economic value of the externalised damage, and consequently the incentives of abating emissions are not altered (Dissou, 2005: 194).

Subsidies, as the third state intervention in environmental issues, can take the form of input, output and emission subsidies. The state subsidises input or output goods that are perceived as being beneficial for emission reductions (Weersink et al., 1998: 317), or the state directly addresses the amount of an installation's pollution and funds every unit not discharged. In a Pigouvian sense, an efficient grant per polluted unit must equal the value (i. e. the social cost) of the connected externality (Migué and Marceau, 1993: 357). Due to the existing external effect, social and firm-level costs differ, and any economic agent receiving the subsidy is incentivised to lower his installation's pollution. Foregoing the benefit implies incurring opportunity costs (Dewees and Sims, 1976: 324), which is why the agent will engage in abating another emission unit as long as the public funding received is greater than the resulting marginal firm-level costs. These incentives of emission subsidies exhibit the same patterns as those of Pigouvian taxes and emission trading schemes, but only in a short-term examination of a competitive market (Goulder and Parry, 2008: 155). In the long run, the intended effect is reversed and aggregate emission and production output rise: subsidies alter the profit structure, and hence the barriers to entry and exit of an industry (Georg et al., 1992: 535). That is why new entrants are attracted due to the cash receipts from the state; and furthermore, subsidised firms are not coerced to internalise any share of the external effect. However the beneficial short-term effects of emission subsidies are very likely not to come into effect since the public regulator is faced with the complex information problem in determining the amount of the external effect (Baumol and Oates, 1975: 135). From an economic perspective, subsidies generally fail in efficiently regulating emissions despite the rule-setting and enforcement competencies of the state (Migué and Marceau, 1993: 360). It can be further doubted that subsidies receive the societal legitimation over a longer period of time: not only can be doubted their mitigation effect on emissions, they also constitute a steady transfer in favour of the polluters – mostly large corporations (Georg et al., 1992: 535).

Taxes are the fourth statist instrument. Designing tax schemes in environmental politics are likely to overcome the widespread criticism towards subsidies. The collection of taxes on output goods[5] is most commonly justified with the Pigouvian hypothesis that the free usage of a resource is efficiently taxed if the levy equals to the social costs of the externality (Pigou, 1962: 192). Charging a polluter in this way changes the alternatives set of his decision space similar to subsidy schemes, and consequently, the emitting activities are lowered to the extent of the firm-specific opportunity cost criterion (Helm and Pearce, 1990: 5). In the Pigouvian sense, the collected cash further leads to internalising the detrimental effects, as long as it is redistributed to the victims compensating for the externality (Baumol and Oates, 1975: 27). On competitive markets, this approach induces economic incentives and solves two issues at once: it provides a reduction of the emissions as well as a Pareto-optimal internalisation by and compensation of the related parties. Because of the information deficit of the tax authority, a broad range of literature discusses the conditions under which an optimal tax rate can be determined (e. g. Carlton and Loury, 1980; Cropper and Oates, 1992). Relaxing the traditional assumption of perfect competition, Barnett (1980: 1040) and Simpson (1995: 364) find for monopolies and oligopolies, respectively, that the regulator can only determine a distorted tax rate. Pigouvian taxes imply a mixed policy mode: on the one hand, the state enacts and enforces

---

5 Apart from this, environmental taxes can be levied to fund direct expenditure programmes or as insurance premium under a trust fund regime (Barthold, 1994: 134-135). Further, taxing relevant input goods can indirectly lead to emission reductions (Goulder and Parry, 2008: 157). However, these types do not aim at internalising an external effect and are largely disregarded in the economic literature due to their hypothesised economic inefficiency.

designated tax laws rendering emissions costly. On the other hand, the way of how this is accomplished by the regulated entities is left to a laissez-faire policy with respect to the choice of means.

The resulting economic incentives differ broadly between the presented state-based environmental regimes. In this respect, technology standards can be used to promote the development of new production processes. Given a competitive market, a tax scheme efficiently lowers emissions and enables the regulator to compensate the victims. Subsidies lack the taxpayers' legitimation and in the long run lead to new entrants to the industry under scrutiny. Apart from these characteristics, all instruments bear one common and prominent shortcoming: the information deficit of the regulator. Especially, the firm-level abatement costs arising from the specific production process cannot be determined. Another aspect is the inflexibility of these measures since the mandated emission reductions cannot be traded among the polluters. This constrains the economic efficiency considering that the regulatory burden cannot be directly balanced between the regulated agents.

2.2.1.3 Negotiations as the Communitarian Approach

Compensation schemes based on negotiations between polluter and victims are advanced by Coase (1960) in order to achieve in an efficient allocation of resources. In the theoretical setting, a compensation solution between the emitting firm and affected parties always leads to an efficient internalisation of the external effect. This outcome does not depend on the ex ante distribution of property rights: either the polluter pays to compensate for his usage of another one's resource, or the victims pay the firm to reduce its emissions (Coase, 1960: 8; Helm and Pearce, 1990: 6). As opposed to state interventions, the purpose is not to attain a predefined and politically desirable emission volume, but instead a direct settlement by means of wealth transfers on a communitarian basis. The role for the state is reduced to reacting on failures within the negotiation scheme: public institutions are only needed in enforcing contracts, preventing abuse by one party and settling disputes (Glaeser et al., 2001: 854).

Simplifying assumptions of this theorem raised major criticism towards the model: Coase (1960: 8) assumes that there are no transaction costs in bargaining the settlement. Further, a negotiation scheme always requires the property rights of the used resource – inducing the commodification of otherwise free goods like air (Quiggin, 1988: 1077). Additionally, full information of either party is required. This hypothesis blanks out all agency conflicts between e. g. a well-informed emitter (agent) opportunistically utilising the asymmetry as to the victims (principals), or vice versa (Alchian and Demsetz, 1972: 787-788; Farrell, 1987: 117; de B. Harris, 1990: 703). Finally, compensation schemes are assumed to take place on a one-to-one basis, but it is more likely to find negotiations between an industry of emitters and the population (or their representatives) of several regions or countries. Transaction costs, information asymmetries and the problem of free riders all increase with the number participants involved (Dixit and Olson, 2000: 312). Especially the latter two shortcomings of the communitarian approach render it unsuitable in applying it to the complex setting like an EU wide emission control.

2.2.1.4 Shift to Markets for Emissions

Emission trading schemes are the most recent policy instrument in environmental policy, and implement the basic idea of Coase (1960) to pollution control as far as possible. Avoiding most of its shortcomings, schemes rely on the framework first derived by Crocker (1966: 81-85), Dales (1968: 801-802; 1970: 93-96) and Montgomery (1972). In environmental policy, credit-based and cap-and-trade schemes are considered.[6] In credit-based systems – similar to performance standards – only emissions beyond or below a legally defined base line are traded. In cap-and-trade systems, the complete amount of an aggregate volume is traded among the participating parties (Tietenberg, 2003: 108). The latter model dominates in the policy-makers' choice, and is the conceptual basis to the EU ETS.

In a cap-and-trade scheme a regulatory authority ex ante defines a maximum volume of certain substances to be emitted during a time period that is allowed by a particular population of emitters. This aggregate volume is subdivided into emission rights representing the property right at the resource used as sink. All regulated polluters have to surrender an amount of certificates equalling their physical emissions after the period has elapsed. Allowances enter the cycle by

6 Crals and Vereeck (2005: 202) further distinguish between average and usage rights schemes. Both have no relevance in environmental policy.

(any possible form of) allocation from the authority, and are traded during the period between polluters, victims or arbitrageurs. The market for emission rights emerges as soon as one entity needs more units than currently at hand while another participant is long. In line with the property rights theory, the price for one certificate equals the underlying external effect, given a fully competitive allowance market is present (Montgomery, 1972: 410). A firm buys emission rights to settle its legal obligation as long as their market price is lower than the specific, process-related abatement cost. In this respect, the prescribed emission volume is attained efficiently, i. e. at marginal costs for the polluters (Lyon, 1982: 18; Labatt and White, 2007: 137-138).

Hence, emission rights provide for an efficient resource allocation, like optimal emission subsidies (in the short run) or optimal Pigouvian taxes do. However, the trade is superior to these and all other purely state-governed instruments. The regulator only has to determine an emission budget and to accomplish an allocation mechanism, relaxing the bounded rationality issue at the authority in this manner (Tietenberg, 1990: 21). Further, only the transfer of rights constitutes an economic compensation between those emitting more and less than the targeted volume – under a Pigouvian scheme, there is merely a relative relief from the tax burden.

An ETS introduces a maximum of laissez-faire into environmental policy, and the functions of the state are reduced to allocating and collecting allowances. To initiate the trade, the regulator transfers the largest part of the budget right away either at the beginning of a longer trading phase or at the beginning of a compliance year, but an even more frequent allocation, i. e. constantly during the trade is also possible (Neuhoff et al., 2006b: 74). Rights can be allocated for free by a grandfathering or benchmarking procedure; they can also be allocated against a fee by means of sales and auctions. Any combination is possible (for a comprehensive overview see Neuhoff et al., 2006a: 419).

The neoclassical assumptions posit that the emission right market always adjusts to an efficient equilibrium, irrespective of the underlying allocation modes or firm characteristics (Montgomery, 1972: 409). Under imperfect conditions these modes play a decisive role if large polluters and product market participants are present (Burtraw et al., 2002). Under free allocation regimes, emission rights are transferred without any monetary equivalent by the polluters. Grandfathering implies a free allocation based on a firm's historic emissions during a previous reference period, resulting in a fraction of business-as-usual emissions for incumbent firms (Åhman et al., 2007: 1719; Ellerman and Buchner, 2008: 71). Since this period is theoretically defined ex post, the regulated entities are unable to bias the physic discharge opportunistically. A repeated grandfathering leads to a moving window of base years and is in the long run subject to emission management concerns (Demailly and Quirion, 2006: 95). Benchmarking, as the second free allocation mode, implements a uniform allocation with respect to a predefined technological standard (i. e. emissions in tonne per output good), irrespective of firm-specific historical emissions (Boemare and Quirion, 2002: 220). Even if opportunistic conduct of the regulated firms cannot be excluded under this approach, it still favours installations with comparably lower emissions by giving rise to ceteris paribus an allowance long position (Burtraw et al., 2002: 52). Since freely allocated rights "only" induce opportunity costs, the risk of adding the full price for emissions on the product price – given no cash outlay occurred – looms large. Such a cost pass-through to customers is usually called reaping windfall profits (cf. Verbruggen, 2008, who refers to "swindle profits"). That is why, costly allocation methods are often justified with the Pigouvian argument of a rent redistribution from polluters back to their victims (Sijm et al., 2006: 50-52). To this end, the regulator can either place certificates at current market price or auction units off. Since the economic consequences of different types of auctions are debated (e. g. Mandell, 2005: 99), safety clauses like price caps are advanced to lower the uncertainties towards future developments of the trading scheme (Grubb and Neuhoff, 2006: 23).

From a business perspective, a cap-and-trade scheme with free allocation creates several typical issues:

1. A certain number of certificates is allocated for free (i. e. represent a grant from the state) while extra certificates have to be purchased either from the regulator or on the market.

2. Emission rights before maturity fulfil the definition of assets and can be purchased or sold at any time on an active market.

3. The obligation emerging by constant emissions is settled in surrendering the respective amount of emission rights (i. e. the liability cannot be settled in cash).

4. Physical emissions induce an economic burden by means of opportunity costs, a purchase of additional rights, investments in pollution abatement measures or a reduction in output.
5. Variation in emission rights prices gives rise to holding (or trading) gains or losses.

If the strict neoclassical assumptions are relaxed, an emission trading scheme can fail in efficiently internalising an external effect. In this respect Lyon (1982: 28) shows that a small number of regulated entities, and hence a small number of emission market participants, leads to distorted or manipulated certificates prices. Further, the presence of an entity with emission right market power impedes the adoption of pollution abating technologies (Malueg, 1989: 55), and leads to increased allowance prices in case it is under-endowed (Hahn, 1984: 756-757). Since suchlike imperfect market conditions are likely, the generally advantageous features of an emission trading scheme over other policies are at risk.

2.2.2 Design of the EU Emission Trading Scheme

2.2.2.1 European Legal Framework

The Kyoto Protocol of 1997 is the first international agreement on climate change and pegs the pollution of six greenhouse gases (Barrett, 1998: 20). Its framework differentiates between developing economies on the one hand and developed as well as transformation countries (so called Annex B parties) on the other. Whereas the former cluster is not subject to any provisions, 37 countries in the latter ratified the document and generally have to comply with its mechanisms – out of them, only the US and Australia opted out of the Protocol (Labatt and White, 2007: 10). The EU-15 Member States agreed to reduce their emissions during the compliance period between 2008 and 2012 by 8% with respect to the base year 1990 (Decision 2002/358/EC:[7] Annex A). In the 2002 EU Burden Sharing Agreement, this aggregate emission target is further specified for

[7] See the Protocol text in Annex I to Council Decision 2002/358/EC concerning the approval, on behalf of the European Community, of the Kyoto Protocol to the United Nations Framework Convention on Climate Change and the joint fulfilment of commitments thereunder, OJ L 130: 1-24.

each Member State. It takes into account present and that time expected specific national emission patterns that are either determined by the energy generation and industry structure or by the projected economic development (Convery and Redmond, 2007: 90). National targets for 2012 as to the base year 1990 range from -28% for Luxemburg and -21% for Denmark and Germany, respectively, till +25% for Greece and +27% for Portugal (Decision 2002/358/EC: Annex II).

Based on the Kyoto emission budgets, an intergovernmental cap-and-trade scheme was installed among the Annex B parties (Annex I, Art. 17). A trade in the so called Assigned Amount Units (AAUs) is heavily disputed since the Central and Eastern European transformation economies were endowed according to their socialist-era emission levels, which have drastically dropped due to industry restructuring (van Vuuren et al., 2006: 446). Since the US did not ratify the Protocol, an excess supply is predicted if these countries attempt to sell their supernumerous units. To further ease attaining the national targets, Annex B countries can generate emission rights additional to the AAUs by means of three flexible mechanisms laid out in the Protocol (Woerdman, 2000: 29-30). Firstly, domestic agricultural and reforestation projects giving rise to sinks for greenhouse gases create Removal Units (RMUs). An amount of emission rights equalling to the emission saving is added to the allocated national cap (Art. 3(4); Dumanski, 2004: 258-259). Secondly, cross-country investments into emission reducing projects domiciled in another industrialised country are supported and termed Joint Implementation (JI). The resulting Emission Reduction Units (ERUs) have the same legal effect like RMUs (Art. 6; Barrett, 1998: 30). Thirdly, in analogy to within-Annex-B-country projects, investments with a pollution reduction effect between industrialised and developing countries are added to the budget. They are called Clean Development Mechanism (CDM) and give rise to Certified Emission Reductions (CERs – Art. 12; Woerdman, 2000: 30-31).

The Kyoto Protocol does not formulate any provisions by which means the industrialised countries are to attain their targets. In 2003, the EU Member States agreed to set up a classical cap-and-trade mechanism that is not linked to the AAU trade of the Protocol and should not mixed up with this instrument. After the regulation had come into effect in 2005, the geographical coverage of the 25 Member States scheme was extended in two steps: With the enlargement of the EU to Bulgaria and Romania, both countries started capping their emissions from 2007 onwards. The three EFTA (European Free Trade Association) members also belonging to the EEA (European Economic Area), Iceland, Liechtenstein and Norway, joined into the scheme in 2008.[8]

Legal basis for the EU ETS is Directive 2003/87/EC,[9] laying down the basic framework with respect to allowed allocation modes (Art. 10-11), to characteristics of allowances (Art. 12-13), to the reporting and verification process (Art. 14-15) as well as to the tasks of national regulators (Art. 16-21). Major features of the ETS – the emission cap, the included industries and the regulated gases – are fixed during so called phases. The pilot period comprised the three years from 2005 until 2007, and the Kyoto compliance period runs for the five year from 2008 until 2012 (Skjærseth and Wettestad, 2008, 2008: 279). The legal framework was amended in 2004 by the EU ETS Linking Directive (Directive 2004/101/EC[10]), whose provisions came into effect for the second trading phase. Its main objective is to open the scheme for the project-based mechanisms of the Kyoto Protocol implying a trade of ERUs and CERs at firm level (Art. 11a). Generally, the EU ETS only regulates carbon emissions of the heavy industries during the first two phases. The regulation leaves the other Kyoto greenhouse gases as well as carbon emissions from households, traffic and further sectors unconsidered. The following activities are affected by Annex I of Directive 2003/87/EC (see also Ellerman and Buchner, 2008: 274):

1. Combustion: Generation of electricity, heat and steam;
2. Mineral oil refineries;
3. Metals: Coke ovens as well as ore, iron and steel processing;

---

8 Even though the EU ETS became a de facto "EEA ETS" in the meantime, the literature still refers to the initial notion due to the limited influence of the three EFTA members on the political arrangements of the scheme. In line with that, this analysis focuses on the European Union and its institutions.

9 Directive 2003/87/EC of the European Parliament and of the Council of 13 October 2003 establishing a scheme for greenhouse gas emission allowance trading within the Community and amending Council Directive 96/61/EC, OJ L 275: 32-46.

10 Directive 2004/101/EC of the European Parliament and of the Council of 27 October 2004 amending Directive 2003/87/EC establishing a scheme for greenhouse gas emission allowance trading within the Community, in respect of the Kyoto Protocol's project mechanisms, OJ L 338: 18-23.

4. Minerals: Production of cement and lime, of ceramics, bricks and tiles as well as of glass;
5. Pulp, paper and board.

As Table 3 shows, the scheme regulates around 2 billion tons of annual carbon emissions of almost 11,000 installations throughout the EU. Presently, it is the largest and the only supranational ETS in force (Neuhoff et al., 2006b: 74). The largest share of the cap is made up among six big countries, namely Germany (23%), and then the UK (13%), Poland (11%), Italy (10%), Spain (8%) and France (7%); the remaining 29% of the cap are subdivided between the other 19 Member States. The striking stake of Germany, the UK and Poland cannot only be attributed to their national economic structure dominated by the heavy industries, but also to operating a considerable pool of coal-fired combustion power plants (Convery et al., 2008: 17).

Since only the aggregate national caps are derived from the EU Burden Sharing Agreement, the specific greenhouse gas emissions have to be defined for all polluting sectors. Further, as opposed to treaty provisions or regulations, European directives generally do not have immediate effect but are to be transposed to national rules in a subsequent process (Börzel, 2001: 805). That is why the broad provisions of the ETS framework have to be broken down to the respective activities and installations in a detailed way. These rules were compiled in respective National Allocation Plans (NAPs) in advance of both first trading phases. Each Member State used its discretion with respect to opt-ins or opt-outs for certain plant or process types, to the definition of industries, to the distribution of the country cap on the dedicated sectors, and to the degree of costly allocation (restricted to 5% in the first and to 10% in the second phase). Due to the relevance of the EU ETS towards the affected industry sectors, the national governments acted strategically and compiled NAPs with favourable stipulations for domestic firms (Åhman et al., 2007: 1726).

Nevertheless, the Commission had the function to review these plans to meet the obliged emission target. Evaluation criteria comprised the freedom from advantages or discrimination for certain industries and from cross-country effects (see the assessment criteria in Directive 2003/87/EC: Annex III; Betz et al., 2004: 376). Table 3 illustrates that the Commission reduced the proposed

Table 3: Scope of the EU ETS

This table comprises the main emission characteristics of all EU-25 member states. The other five participants in the scheme (Bulgaria and Romania as EU-27 Member States; Iceland, Liechtenstein and Norway as the remaining three members of the EEA) are omitted for the sake of comparability: They joined the scheme in 2007 and 2008, respectively, and the NAPs of the three non-EU members are assessed by the EFTA Surveillance Authority. Figures for the EU ETS Phase I (2005-2007) and Phase II (2008-2012) as indicated. Data taken from: EU press releases IP/04/862, IP/04/1250, IP/05/9, IP/05/269, IP/05/422, IP/05/602, IP/05/762, IP/07/1566 and IP/09/794.

Member State	Cap allowed[a]		Share of cap[b]	Installa-tions[c]	Proposed vs. al-lowed cap[d]	
	Phase I	Phase II	Phase II	2008	Phase I	Phase II
Austria	33.0	30.7	0.0157	216	0.9970	0.9360
Belgium	62.9	58.5	0.0298	302	0.9890	0.9242
Cyprus	5.7	5.5	0.0028	13	1.0000	0.7697
Czech Rep.	97.6	86.8	0.0442	401	0.9047	0.8518
Denmark	33.5	24.5	0.0125	378	1.0000	1.0000
Estonia	19.0	12.7	0.0065	50	0.8753	0.5217
Finland	156.5	37.6	0.0192	600	1.0000	0.9495
France	45.5	132.8	0.0677	1,016	0.9681	1.0000
Germany	499.0	453.1	0.2310	1,668	1.0000	0.9400
Greece	74.4	69.1	0.0352	139	1.0000	0.9152
Hungary	31.3	26.9	0.0137	237	1.0000	0.8762
Ireland	22.3	21.2	0.0108	105	0.9993	0.9358
Italy	232.5	195.8	0.0998	1,048	0.9100	0.9368
Latvia	4.6	3.3	0.0017	86	0.7098	0.4286
Lithuania	12.3	8.8	0.0045	111	0.9042	0.5301
Luxembourg	3.4	2.7	0.0014	15	0.8776	0.6835
Malta	2.9	2.1	0.0011	2	1.0000	0.7095
Netherlands	95.3	85.8	0.0437	375	0.9695	0.9491
Poland	239.1	208.5	0.1063	858	0.8354	0.7326
Portugal	38.2	34.8	0.0177	213	0.9820	0.9694
Slovakia	30.5	30.9	0.0158	179	0.8600	0.7482
Slovenia	8.8	8.3	0.0042	91	1.0000	1.0000
Spain	174.4	152.3	0.0776	1,039	1.0000	0.9974
Sweden	22.9	22.8	0.0116	770	1.0000	0.9048
Un. Kingdom	245.3	246.2	0.1255	952	1.0000	1.0000
Sum: EU-25	2,190.8	1,961.7	1.0000	10,864	0.9573	0.9073

[a] In million tons of carbon annually, as approved after the review process.
[b] Expressed as national emission cap in relation to EU-25 emission cap.
[c] Number of installations; definition of the term installation varies between a facility and a plant among the Member States.
[d] Expressed as proposed cap in NAP first submitted to the review in relation to the accepted cap.

caps by around 4% for the first (14 NAPs corrected) and by 9% for the second trading period (21 Member States affected). Whereas the Central and Eastern European countries' budgets were more thoroughly reduced (e. g. for Phase II: Slovakia to 75%, Poland to 73% as well as the Baltic countries to 43%-53%), most Western European countries were able to pursue their goals with only small changes. Usually, the British NAP is regarded as the most restrictive and is hence considered as model in an ecological sense, compared e. g. to the German plans that were perceived to be driven by the industry's interests (Endres and Ohl, 2005: 26; Skjærseth and Wettestad, 2008: 280-281).

Once the emission budget is determined, its freely available contingent is allocated to the regulated firms on February 28 on a yearly basis (Directive 2003/87/EC: Art. 11(4); Labatt and White, 2007145). Various exchanges across Europe quote spot, futures and option contracts daily. The most liquid exchanges are European Energy Exchange (EEX, Leipzig), BlueNext (Paris; previously part of PowerNext), European Climate Exchange (ECX, London), NordPool (Lysaker/Oslo), Energy Exchange Austria (EXAA, Vienna) and Climex (Utrecht). Generally, the annual cycle of compliance and trading is shifted by some months. Since emission volumes are measured for the twelve months of a calendar year, the verification takes place at the beginning of the consecutive year. Until the subsequent March 31, an emission report on the number of allocated and verified emissions with respect to the preceding calendar year is to be submitted at the national regulator (Directive 2003/87/EC: Art. 14(3)). In accordance with this report, allowances have to be surrendered by April 30 when the legal settlement of the obligation, that is the voiding of the emission rights, occurs (Directive 2003/87/EC: Art. 12(3); Labatt and White, 2007: 145). The filings of the national registries are aggregated in an EU database, the Community Independent Transaction Log (CITL). Shortly after the end of the annual trading cycle, the Commission publishes by this means annual reports on the past regulatory outcome and on the allocation states of every installation.[11] Hence, the trade in allowances for settling the prior year obligation also stretches to the first four months of a following year, as presented in the time line of Figure 2.

---

[11] All documents are available via the internet platform of CITL at: http://ec.europa.eu/environment/climat/emission/citl_en.htm.

Figure 2: Emission Trade Calendar
(Own illustration based on BMU, 2007: 33)

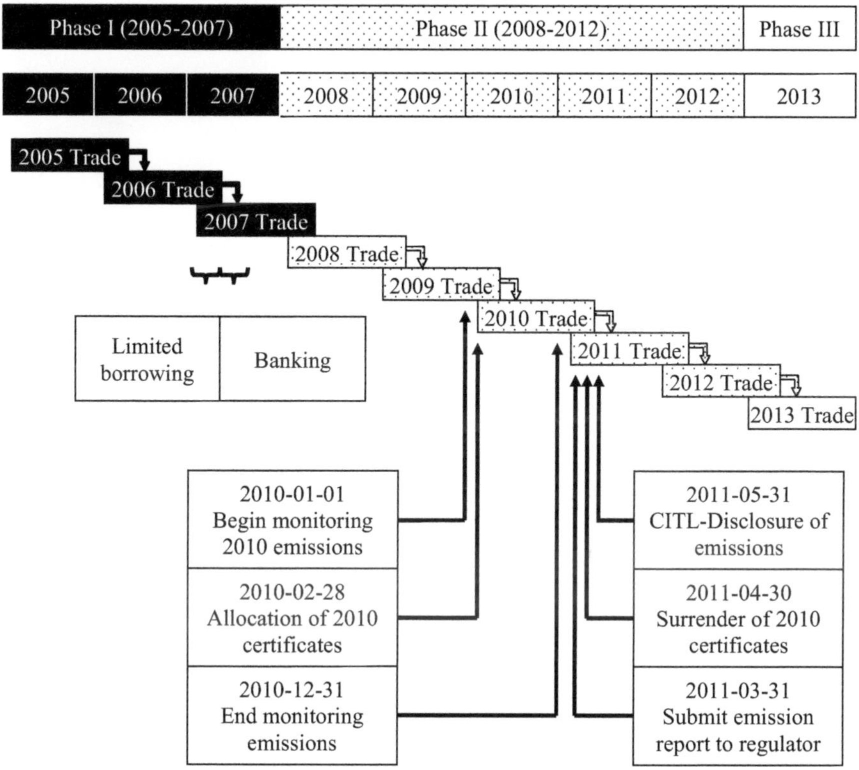

If an entity does not dispose of the required amount of emission rights at the end of a compliance cycle, it is obliged to surrender the missing allowances one year later. This non-compliance is disclosed and can be accessed by anyone in the CITL database. Additionally, a shortfall of available certificates for surrender is fined by 100 Euro (by 40 Euro during the first phase between 2005 and 2007) per tonne (Directive 2003/87/EC: Art. 16, paras. 3-4; Kruger and Pizer, 2004: 22). These provisions support the incentive to either reduce own emissions or to level a potential shortfall in time before the end of the trade by purchasing certificates from other market participants. In addition, short and long positions can be balanced at installation level. By means of banking, unused emission rights can be transferred to a consecutive year (this was not possible between the first and second trading phases). The reverse transaction, i. e. borrowing allowances from a preceding year, is only possible in a restricted way since the allocation for the present year's rights precedes the surrender for the elapsed year by one month (Convery and Redmond, 2007: 96).

2.2.2.2 National Allocation Plans: The German Case

Two sources determined national legal actions of the EU ETS during the first two phases: the persistent, cross-phase emission directive, and the transitory, phase-related rules of the reviewed NAPs. Both requirements had to be introduced into national legislation to become effective for the market participants. Brief reviews and comparisons of the first and second phase plans can be found in DEHst (2005) and Rogge et al. (2006), respectively. Germany's NAP serves as an exemplary case in this study due to its prominence within the EU ETS. The underlying mechanisms are similar for all other EU member states.

The federal German government adopted the general purpose Greenhouse Gas Emission Trading Act (Gesetz über den Handel mit Berechtigungen zur Emission von Treibhausgasen – TEHG) that basically transposed Directive 2003/87/EC. Allocation Acts (Gesetze über den nationalen Zuteilungsplan für Treibhausgas-Emissionsberechtigungen – ZuG) and additional regulations reflect the provisions of the NAP for a designated period. Technically, the lawmaker chose a two tier approach to determine the respective volume of emission rights granted free of charge. On the first tier, the so called macro plan, the maximum budget available to the regulated industries was derived from the country emission target (Betz et al., 2004: 379-380). On the second tier, the micro plan, the allocation modes were prescribed.

The micro plans generally distinguished between industries with process-related and energy conversion emissions. In the former case, emissions arise due to a chemical reaction in the course of the production, e. g. in processing cement or iron ore, and cannot be easily abated. The latter case, energy conversion emissions, can be found in all fossil power plants or in the manufacture of bricks and stoneware (El-Fadel et al., 2001: 1038). For the first phase, the whole national budget was allocated by means of grandfathering, i. e. the number of free allowances to obtain yearly was computed as a fraction of emissions during certain base years, and yielded a so called compliance factor. The general first phase compliance factor was set to 95.38% of the base years (DEHSt, 2004: 2), but some installations like firms with large volumes of process-related emissions were partly or completely exempted from reduction provisions (Burgi and Lange, 2006: 545).

Starting in 2008, the second phase witnessed some revisions to the legal basis. In reaction to the 2004 amendment of the emission trading directive, the new rules specify the amount of project-based Kyoto allowances eligible for surrender. According to this provision, up to 22% of the allocated emissions may be surrendered in form of CERs and ERUs – the rest can only be settled by means of EU allowances (ZUG 2012:[12] § 18). Additionally, the procedures underlying the free allocation were changed: incumbent installations with process-related emissions remained under the grandfathering mode equalling to an allotment of 98.75% of historical emissions (§ 6(1)). Installations with process-related emissions having entered into service after 2003 as well as all electricity generating entities were allocated according to a best available technology benchmark, expressed in grams of carbon emitted per output unit (§§ 7(1), 8(1)). Changing the allocation modes especially to the detriment of the energy industry was a reaction of the federal government to the investigations by the German competition authority, finding windfall profits and a (an opportunity) cost pass-through to power customers (Bundeskartellamt, 2006: 6).

Around 8.8% of the yearly emission cap (equalling to 40,000,000 certificates) were reserved for selling off, being close to the maximum of 10% of national costly allocated rights (Directive 2003/87/EC: Art. 10). These allowances have been offered by the state-owned banking group KfW in small tranches at the ECX and EEX at market value throughout the annual cycle (BMU, 2008).

During the remaining three years of the second phase, it is planned that the number of emission rights allocated for money will remain unaltered, but the mode of disposal will be changed to weekly auctions on emission exchanges (ZuG 2012: § 21(1); EHVV[13]: § 2-3). The employed uniform price sealed bid auction method is supported by economic theory. It provides for maximum

---

12 Gesetz zur Änderung der Rechtsgrundlagen zum Emissionshandel im Hinblick auf die Zuteilungsperiode 2008-2012, BGBl I 38 (2007), 1788-1808.

13 The bill for the Emission Trade Auction Regulation (Verordnung über die Versteigerung von Emissionsberechtigungen nach dem Zuteilungsgesetz 2012 – EHVV) is to be ratified by the German Lower House, see Drucksache des Deutschen Bundestags 16/13189 of 2009-05-27.

revenues since bidders pay the marginal price on auctions of homogenous goods like emission allowances (Franciosi et al., 1993: 12).

2.2.2.3 Supply of and Demand for Emission Rights

The pilot phase of the EU ETS witnessed a debate on the systematic over-endowment (cf. Ellerman and Buchner, 2008: 268). Calculating the relation between allocated emission rights and verified emissions yields the national allocation status for each year as displayed in Table 4. It is apparent that the collective first phase allocation started at 109% in 2005 and ranged at 107% in 2007. Remarkable excessive supplies can be found foremost in the Baltic countries (e. g. Lithuania: 186%-205%), but also in large Member States (e. g. France: 119%-124%; Netherlands: 119%-124%; Poland: 114%-118%). Consecutive shortfalls were only reported in Slovenia (97%-99%), Spain (93%-97%) and the UK (96%-98%). The fact that nearly every participating country exhibited less pollution than allowed raised the question whether the Member States were interested in a strict and functioning scheme (Skjærseth and Wettestad, 2008: 284). Another major concern did not address the quantity of the budget but its mode of allocation: all first phase NAPs relied on granting emission rights either by grandfathering or by benchmarks (DEHSt, 2005: 16-19). Consequently, some economic studies found evidence for an emission cost pass-through or even ETS-induced windfall profits (Sijm et al., 2006: 61-67).

To overcome both issues, the Commission announced to enforce more harmonised allocation modes and a stricter cap for the Kyoto compliance period (Convery et al., 2008: 10). The endowment figures for 2008, the first trading year of the second phase, back this assertion. This policy change affected an aggregate shortage of the resource air, with an EU-wide allocation status of 98%. Further, eleven of 25 countries were short, among them – except for Poland – all major emitters: Germany (96%), Spain (93%), the UK (93%) and Italy (89%). Additionally, after all Member States had refrained from costly allocation modes in the first period, Britain and Germany introduced auctions and sales in 2008. Austria, Ireland and the Netherlands followed in 2009. However, only Britain and Germany are close to or at the 10% threshold, the others announced to sell

Table 4: Allocation of Emission Rights: Country Perspectives

This table comprises the main emission characteristics of all EU-25 member states. The other five participants in the scheme (Bulgaria and Romania as EU-27 Member States; Iceland, Liechtenstein and Norway as the remaining three members of the EEA) are omitted for the sake of comparability: They joined the scheme in 2007 and 2008, respectively, and the NAPs of the three non-EU members are assessed by the EFTA Surveillance Authority. All figures presented are on an annual basis. Data taken from: EU press releases IP/07/1566 and IP/08/787; European Commission (2009), DMO (2009).

Member State	Allocation status[a]				Share of costly allocation[b]	
	2005	2006	2007	2008	2008	2009
Austria	0.9888	1.0191	1.0393	0.9593		0.0130
Belgium	1.1367	1.1489	1.1920	1.0547		
Cyprus	1.1144	1.0762	1.0489	0.9777		
Czech Rep.	1.1837	1.1671	1.1112	1.0840		
Denmark	1.2653	0.9795	1.1392	0.9230		
Estonia	1.5014	1.5649	1.2361	0.9391		
Finland	1.3746	1.0197	1.0695	1.0398		
France	1.1923	1.2325	1.2358	1.0758		
Germany	1.0505	1.0439	1.0243	0.9587	0.0883	0.0883
Greece	1.0440	1.0634	1.0231	0.9892		
Hungary	1.1951	1.2097	1.1651	0.9873		
Ireland	0.9952	1.0289	1.0512	1.0377		0.0263
Italy	1.0288	1.0223	1.0270	0.8873		
Latvia	1.5998	1.5529	1.6028	1.2028		
Lithuania	1.8575	1.8823	2.0449	1.4417		
Luxembourg	1.2894	1.2373	1.3075	1.2864		
Malta	:[c]	:[c]	1.4518	1.0403		
Netherlands	1.1860	1.2425	1.1931	1.0274		0.0373
Poland	1.1770	1.1407	1.1406	1.0215		
Portugal	1.0478	1.1536	1.2221	1.1633		
Slovakia	1.2088	1.1941	1.2440	1.2123		
Slovenia	1.0053	0.9915	0.9688	0.9368		
Spain	0.9499	0.9706	0.9349	0.9317		
Sweden	1.1815	1.1517	1.2027	1.1396		
Un. Kingdom	1.0116	0.9768	0.9562	0.9289	0.0162	0.1000
Sum: EU-25	1.0888	1.0773	1.0656	0.9823	0.0224	0.0351

[a] Expressed as emission rights allocated in relation to verified emissions.

[b] Expressed as emission rights sold or auctioned off in relation to allowed cap. Figures for 2009 as proposed by the Member States.

[c] No data available in the CITL database.

only small volumes of allowances. On the aggregate, the share of costly allocation therefore remains for the EU at 2% and 4% for the trading years 2008 and 2009, respectively.

In order to assess the participating countries' ability of pursuing own industrial politics, it can be insightful to focus on the different sectors involved. The underlying industry characteristics – i. e. the market power and the output demand elasticity, the share of imports and exports as well as technical issues – play an important role in both ex ante determining the allowed cap and allowance demand patterns (Bleischwitz et al., 2007: 12). If it is the aim of a policy maker to avoid distorting spill-overs in case of market dominance or competitive disadvantages for export-oriented firms, these installations have to be allocated according to distinguished modes. A frequently used indicator for output market power is the cumulative market share of its largest firms. The higher the so called concentration ratio of x suppliers CR_x, the less likely is competition (Bain, 1951: 311-313). Common empirically determined critical thresholds of a four-firm concentration ratio CR_4 are 40% for monopolistic competition and 60%-70% for an oligopoly (Dalton and Penn, 1976: 140; Weiss, 1979: 1110; Geithman et al., 1981: 351). For the five regulated activities, these economic determinants of are briefly sketched in the following:[14]

- The European *energy market* was for a long time dominated by regional, mostly state-owned mono- or oligopolies. After efforts of liberalisation, several subsequent waves of cross-border and domestic mergers took place (Verde, 2008: 1126). The national markets have remained isolated and play the most important role in political and economic evaluation (Convery et al., 2008: 18). Here, the national CR_3 ranges throughout Europe from 37% in the UK via around 60% – e. g. Denmark, Germany and the Netherlands – to nearly a perfect monopoly with 90% – e. g. Belgium, France and Greece (European Commission, 2008: 12-13).
- European *mineral oil processing* is accomplished within an oligopoly that is dominated by eight large, multinational corporations (Thomas, 2003:

[14] Calculating concentration ratios is complex as it involves the definition of (sub-) industries and their products, the regional extension of a market, as well as the point in time of measurement (Bain, 1951: 298-304). That is why this question is excluded and a potential starting point for future research. In order to come to an economic evaluation, the results of prior contributions are referred to as proxy.

394). Their refining capacity is added by a dozen of (partly former) state-owned entities with regional monopolies. Side-by-side with the energy sector, this is the least competitive ETS industry.

- *Iron and steel smelters* face considerable competition from outside the EU, and hence outside the ETS (Demailly and Quirion, 2008: 2018). High capital intensity and high barriers to entry render the European CR_8 at 57%, so that the industry is characterised as an oligopoly (Oberender and Rüter, 1993: 72-73).
- Most *mineral processing*, like cement, clinker, and concrete, is provided in regional oligopolies due to a limited transportability of these produces (Demailly and Quirion, 2006: 98-99). In most Member States, especially the cement industry is highly concentrated, with national CR_3 around 50% (Wagner and Vassilopoulos, 2000: C1).
- The *pulp and paper industry* has been undergoing a concentration process during the recent decade. Rising capital intensity and the participation in the globalised markets promoted the rise of large corporations. Still, the European industry is dominated by private entities, and a can be termed as monopolistic competition with a CR_{10} of 48% (Bergman et al., 2002: 31-32; CEPI, 2006).

Table 5 summarises the allocation properties of these five sectors (plus certain national opted-in entities) regulated by the EU ETS during the first two phases. It can be observed that combustion installations account for the largest stake by far in the allocation with roughly two third. The metals and minerals processing follow with 12% each. In sum, the positive correlation between industry competitiveness and over-endowment is apparent. The paper and metals industries are the most exposed to imports from firms outside the scheme, and they were consistently long at around 120% to 133%. This furnishes evidence that national allocation rules are being tailored to back these industries. In both cases the risk of a cost pass-through to other market participants is small (Bleischwitz et al., 2007: 16). As highly oligopolised sectors, minerals and oil processing show a weaker over-endowment, and the combustion installations mostly operated by power suppliers were short in three trading cycles – in 2008 even at 83%. On a cautionary note, the implications of analysing the allocation status of Member States and industries should not be over-interpreted. An excess in rights must not necessarily be completely due to a generous allocation, but can also be driven by early abatement measures or by a decline in production, and hence in emissions (Ellerman and Buchner, 2008: 270-271).

Table 5: Allocation of Emission Rights: Industry Implications

Data obtained from the European Environment Agency EU ETS data viewer.

Activity	Allocation status[a]				Share of sector[b]
	2005	2006	2007	2008	2008
Combustion installations	1.0067	0.9834	0.9776	0.8326	0.6522
Mineral oil refineries	1.0539	1.0600	1.0577	0.9782	0.0790
Metals	1.2415	1.1860	1.1815	1.3254	0.1184
Minerals	1.0870	1.0602	1.0368	1.1212	0.1294
Pulp, paper and board	1.2320	1.2343	1.2800	1.2024	0.0202
Other activity opted-in	1.0891	1.1245	1.1597	1.0943	0.0009

[a] Expressed as emission rights allocated in relation to verified emissions.
[b] Expressed as emission rights allocated to a respective sector in relation to total allocation.

The final stage of a general inspection concerning supply and demand patterns of emission rights focuses on the market price developments. Figure 3 displays the quotations for spot and futures contracts from the beginning of the trade until the end of 2009. For reasons of clarity, it only includes a selection of the broad palette of yearly and monthly maturing contracts. Untabulated results of correlation analyses separately conducted for Phase I (i. e. futures maturing in 2005-12, 2006-12 and 2007-12) and Phase II (i. e. futures maturing in 2008-12 through 2013-12) indicate pairwise coefficients above 0.98, statistically significant at the 1% level.

Generally, allowance market developments can be explained by demand side (technical and economic issues) as well as supply side (regulatory) characteristics. Technical issues especially pertain to electricity generation. Price fluctuations in alternative fossil fuels result in lower carbon emissions – for example, a rising price of coal compared to gas alters the power producers' preferences towards the input goods (Paolella and Taschini, 2008: 2023), and changing weather conditions determine the demand for power, and hence allowances (Mansanet-Bataller et al., 2007: 75; Benz and Trück, 2009: 6). Economic issues as to emission rights pricing can arise if firms have market power for allow-

Figure 3: Development of Allowance Prices

EUA Spot: Emission allowances spot contracts, data obtained from European Energy Exchange for the first period and from BlueNext for the second periods; *CER Spot:* Certified Emission Reduction, data obtained from BlueNext; *EUA Futures Dec 200x:* Emission allowances futures contracts mature in 2008-12, 2012-12 or 2013-12, respectively, data obtained from European Climate Exchange.

Panel A: First Trading Period

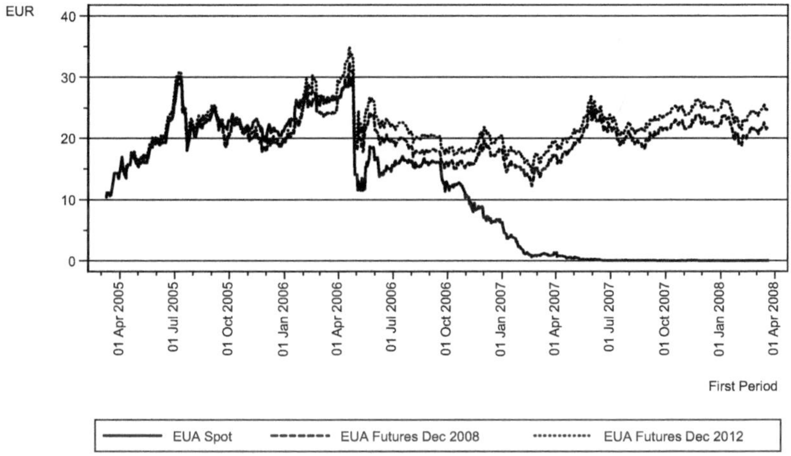

Panel B: Second Trading Period

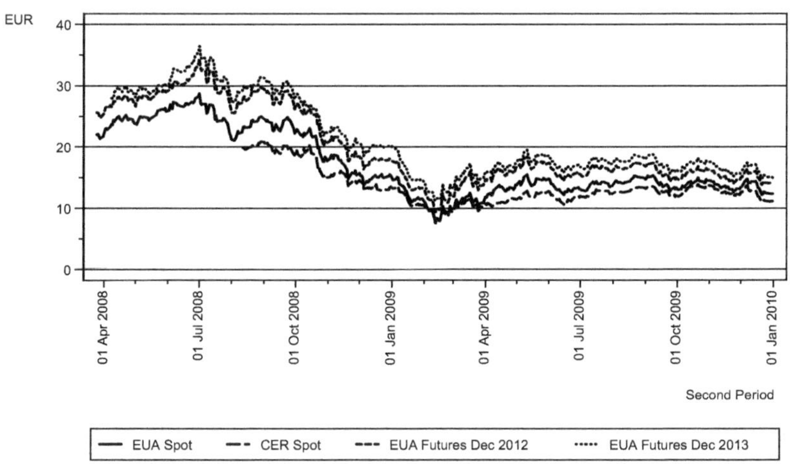

ances. This was not the case during the first phase as the CR_{10} for allocated emission rights stood at 21% and did not indicate low competition on the allowance market (Convery and Redmond, 2007: 102). Far more prominent loomed regulatory issues, especially during the first trading phase of the ETS (Alberola and Chevallier, 2007: 12-13). After the EU Commission had released the first allocation and compliance details in 2006-05, the spot contracts plummeted from almost 30 to 12 EUR/t within a few days. After the price had recovered for several months to around 15 EUR/t, it finally declined to near zero after 2006-10 since it was not possible to bank the first phase spot contracts to the subsequent period (Parsons et al., 2009: 82). The futures predicted at that time a more rigorous regulatory environment for the second phase: contracts maturing in 2008 and 2012 oscillated slightly above their opening price between 20 and 25 EUR/t. In the second period, no decline after the 2008-05 and 2009-05 ETS details publications could be witnessed; in this respect, the importance of regulatory decisions seems to have decreased. The quotations for spot and futures allowance contracts fluctuated around a stable 12 EUR/t at the end of 2009. Spot CERs, that are quoted since 2008-08 and can be used (to a certain extent) to settle the emission obligation, exhibit a stable spread towards the allowance spot contracts.

2.2.2.4 Outlook on the Future Design of the EU ETS

The regulatory environment of the last trading year of Phase II (2012) as well of the whole Phase III (starting in 2013) is presently still diffuse, but it will take shape by the end of 2010. Three documents, Directive 2008/101/EC,[15] Regulation (EC) No. 219/2009[16] (Annex, para. 3.6) and Directive 2009/29/EC,[17] prescribe a number of substantial changes to the current rule set. The most striking

[15] Directive 2008/101/EC of the European Parliament and of the Council of 19 November 2008 amending Directive 2003/87/EC so as to include aviation activities in the scheme for greenhouse gas emission allowance trading within the Community, OJ L 8: 3-18.

[16] Regulation (EC) No 219/2009 of the European Parliament and of the Council of 11 March 2009 adapting a number of instruments subject to the procedure referred to in Article 251 of the Treaty to Council Decision 1999/468/EC with regard to the regulatory procedure with scrutiny, OJ L 87: 109-154.

[17] Directive 2009/29/EC of the European Parliament and of the Council of 23 April 2009 amending Directive 2003/87/EC so as to improve and extend the greenhouse gas emission allowance trading scheme of the Community, OJ L 140: 63-87.

new features of the ETS can be summarised as follows: contrary to initial plans, the third trading period will run for the eight years between 2013 and 2020. In order to be in line with the so called “20-20-by-2020” targets adopted under the EU action against climate change, the aggregate cap for the presently regulated industries will be reduced annually by 1.74% to achieve an emission reduction of 21% by 2020 compared to 1990 levels (Art. 9;[18] Council of the EU, 2009: 3). Grandfathering as mode for a free allocation of allowances will be completely replaced by benchmarking in all sectors, developed as a standard with respect to the performance of the 10% most efficient entities (Art. 10(2)). The allocation provisions will no longer be set out in NAPs but directly by the Commission (Art. 10(1) and 10a(1)).

Apart from the previous activities of the heavy industries, the production of aluminium, non-ferrous metals, carbon black, acids, bulk organic chemicals (by e. g. cracking) and other defined chemical substances will be included into the EU ETS. Since some of the named processes also emit the greenhouse gases perfluorocarbons and nitrous oxide (NO_x), the scope of the trade will no longer be restricted to carbon only. Auctioning will gain an ever growing stake of allocation for industry: in 2013, 20% of the allowed cap will be auctioned, steadily increasing to 70% in 2020, and to 100% by 2027 (Council of the EU, 2009: 3). Even though the responsibilities for determining the cap will move to the EU Commission, the Member States retain discretion in defining own auctioning rules (van Zeben, 2009). In sum, the free allocation in 2020 will drop to 23.7% of 1990 emissions levels. Exemptions from costly allocation still have to be determined, but will be limited to address carbon leakage, i. e. to avoid a free-riding effect by displacing regulated activities to outside-ETS countries (Felder and Rutherford, 1993: 163). Firms with foreign revenues greater than 10% and a carbon trade-related production cost increase by at least 5% will be granted a volume of emission rights according to the effective performance benchmark and may additionally receive financial aid by national governments (Art. 10a(6) and (15)). Energy generation will face the same reduction in its carbon emission cap, but the allocation mode is stricter: to address the issue of windfall profits, the complete emission budget will be auctioned off from 2013 onwards (Art.

[18] References to the emission trading Directive within this section refer to the consolidated text as published on 2009-06-25.

10a(1)). To take economic differences among the Member States into account, ten Central and Eastern European countries can apply for reducing auctioning to 30% for incumbent fossil fuel fired plants until 2020 (Art. 10c).

Already from 2012, the scope of the scheme will be extended to carbon emissions by aviation, covering all flights from, to and inside of the EU (see Annex I). Special provisions apply for this industry: the collective cap is fixed as compliance factor based on average historical emissions between 2004 and 2006 (Luhmann, 2008: 232), equalling to 97% for 2012 and to 95% from 2013 onwards (Art. 3c(1) and (2)). 85% of the budget will be allocated for free according to a benchmark (in t CO_2/km), and the remaining 15% will be auctioned off right away from the beginning.

Table 6: Future Allowed Emission Volumes

Sector	Allocation	Phase II	Phase III				
		2012	2013	2014	...	2019	2020
Industry[a]	Sector cap[b]	see NAP	0.9118	0.8944	...	0.8074	0.7900
	Auctioning[c]	≤ 0.1000	0.2000	0.2714	...	0.6286	0.7000
	Free budget[d]	see NAP	0.7294	0.6517	...	0.2999	0.2370
Energy	Sector cap[b]	see NAP	0.9118	0.8944	...	0.8074	0.7900
	Auctioning[c]	≤ 0.1000	1.0000	1.0000	...	1.0000	1.0000
	Free budget[d]	see NAP	0.0000	0.0000	...	0.0000	0.0000
Aviation	Sector cap[e]	0.9700	0.9500	0.9500	...	0.9500	0.9500
	Auctioning[c]	0.1500	0.1500	0.1500	...	0.1500	0.1500
	Free budget[f]	0.8245	0.8075	0.8075	...	0.8075	0.8075

[a] This sector comprises all non-energy and non-aviation activities.
[b] Collective EU-wide cap expressed as fraction of 1990 emissions.
[c] Auctioned share of emission cap.
[d] Volume of certificates allocated for free expressed as fraction of 1990 emissions.
[e] Collective EU-wide cap expressed as fraction of average historical emissions between 2004 and 2006.
[f] Volume of certificates allocated for free expressed as fraction of average historical emissions between 2004 and 2006.

The development of the sectoral caps, auctioning and free budgets is summarised in Table 6. The revenues from auctioning emission rights by the Member States cannot be used freely (Art. 10(2) and (3)). Out of the cash inflow,

12% are redistributed to Central and Eastern European countries, another 50% should be used for defined national climate protection policies.

2.2.3 New Governance Modes for Pollution Control: An Explanation

From a governance perspective two questions emerge: Why was a policy setting chosen in which other actors than the nation state played the most important role? And why was a market-based mechanism preferred to a state- or community-based one? The questions will be addressed in turn. A de-nationalisation of environmental politics is due to the globalised character of greenhouse gas emissions. In the beginnings of environmental policy, local issues like the discharge of smoke or toxic substances were in its focus (see the examples of the early 19th century Manchester in Douglas et al., 2002: 241). Cross-boundary pollution control only entered the political agenda in the 1950s (see for the foundation of the International Commission for the Protection of the Rhine against Pollution in 1950 in Bernauer and Moser, 1996: 390). As greenhouse gas emissions give rise to global externalities, political actions against it cannot be restricted to local or intergovernmental levels; a globalised phenomenon is to be addressed.

The influence of lobby groups in politicising greenhouse gas emissions plays a further role. Lobby groups act as both an intellectual cause in spinning the public opinion, and as institutional cause directly approaching the policy maker (Biliouri, 1999: 174-175). Both causes urged policymakers to debate carbon emissions and other greenhouse gases, and they did so at an international level. International agreements however, resulted from institutional arrangements like the United Nations (Callon, 2009: 539). In the UN, two organisations support and promote the political process of regulating emissions at a global level. This is the Intergovernmental Panel on Climate Change (IPCC), which was established in 1998 to supply policymakers with scientific results (Siebenhüner, 2003: 117). Its assessment reports (e. g. IPCC, 2001; IPCC, 2007) are the basis to formulating and justifying pollution caps or benchmarks. Before, the United Nations Framework Convention on Climate Change (UNFCCC) was launched in 1992 as political forum to reach an international emission reduction commitment. Whereas its early conventions on climate change had only resulted in voluntary agreements, during its 1997 conference in Kyoto the first binding treaty was negotiated (Labatt and White, 2007: 10). Many developed countries, and in particular the EU-15, ratified the Kyoto Protocol and subjected themselves to greenhouse gas and especially carbon pollution abatement measures. The EU-15 was likely to agree as it had established some own environmental policy fields beforehand (Pollack, 1994: 104), and therefore at that time already exerted appropriate influence as institution. The Member States had already learnt that agreeing on the abatement of global greenhouse gas emissions is a tedious political process – too complex for the limited scope of national actors on their own, and again too complex for reaching a detailed statist regulation at the international, intergovernmental level. Due to both, the globalised and complex nature of these emissions, national actors fail in adopting effective measures (Biermann and Dingwerth, 2004: 11). The underlying driving forces of policymaking for emissions are illustrated in Figure 4; these emerging factors promoted a situation in which traditional instruments of national politics had lost their relevance.

Figure 4: Causality Model Explaining Environmental Policy Choice

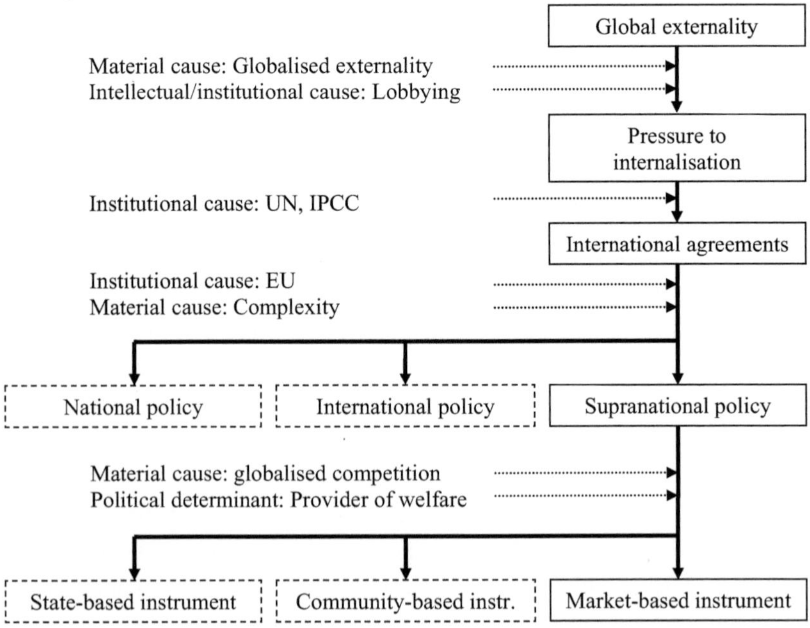

The second question is why the EU chose to implement a market-based instead of state- or community-based policy instruments. Here, it can be observed that certain kinds of state-based regulations dominated for a long time and still play a large role in European environmental politics. Technology standards are e. g. employed in restricting the pollution of volatile organic compounds. If a certain threshold of yearly pollution is exceeded, special emission limits apply (Directive 1999/13/EC:[19] Annex IIA). Also do performance standards serve as carbon pollution control in specific limits for passenger cars, as from 2012 onwards a new vehicle must not emit more than around 130 g CO_2/km, depending on its mass (Regulation (EC) No. 449/2009:[20] Art. 4 in conjunction with Annex I). Fiscal measures are unknown to environmental politics at the European level. Instead, subsidies are employed in national rules like in grants e. g. supporting the research on, development of and generation of power from bio-fuels (Schneider and McCarl, 2003: 293). The same applies to Pigouvian taxes: as a reaction to the Kyoto agreements, the Scandinavian countries and the Netherlands introduced national carbon tax schemes for industrial emitters (Cansier and Krumm, 1997: 65-68). Backed by these moves, the Commission advanced plans in 1992 and 1995 to adopt an EU-wide carbon tax that were rejected by the Member States, though (Convery and Redmond, 2007: 88).

Community-based compensation instruments have not yet been implemented in the pure Coasian sense. Negotiated agreements on e. g. standards between governmental bodies and industrial associations are a common instrument only in German environmental politics and are rarely used elsewhere (Jordan et al., 2003: 209-212). Market-based solutions were first adopted in the US where several emission trading schemes have been adopted in parallel. The first of these mechanisms, the Emission Reduction Credit programme, was introduced in 1976 and worked as a small credit trading scheme based on technical benchmarks (Tietenberg, 1990: 18; Lohmann, 2009: 504). In 1994, California started the Regional Clean Air Incentives Market (RECLAIM), a system only covering

19 Council Directive 1999/13/EC on the limitation of emissions of volatile organic compounds due to the use of organic solvents in certain activities and installations, OJ L 85: 1-22.

20 Regulation (EC) No. 449/2009 of the European Parliament and of the Council of 23 April 2009 setting emission performance standards for new passenger cars as part of the Community's integrated approach to reduce CO2 emissions from light-duty vehicles, OJ L 140: 1-15.

nitrous oxide emissions in its southern regions (Lejano and Hirose, 2005: 369). One year later, the so called Acid Rain Program,[21] began regulating sulphur dioxide (SO_2) in several states on a larger scale (Joskow and Schmalensee, 1998: 40-41). Another trade for nitrous oxide was set up by the Ozone Transport Commission (OTC) programme in several north eastern states in 1999 that was extended to the larger Nitrogen Oxides Budget Trading Program (NBP) in 2004 (Burtraw et al., 2005: 271). Before the EU ETS was decided on, two European countries introduced national trading schemes, both for carbon emissions: the Danish regulation started in 2000, and the British one in 2002 (Pedersen, 2001: 224; Nye and Owens, 2008: 2). Both trades as well as national carbon taxes were abolished as soon as the EU ETS came into effect in 2005.

The emerging question is why the EU Member States did not embark collectively on a policy with a high degree of legitimation among the regulated industries and the population. In this respect, EU level technology and performance standards have already proven to effectively reduce pollution, and could have been a legitimated model for the greenhouse gas emissions case. Two issues played a significant role in choosing another policy instrument: The global competition between countries with their conjoint economic interests of occupying the national workforce or securing future tax collections as well as the global competition between firms situated in these countries which advocated for internationally level playing fields in regulation. Debating this issue of carbon leakage most commonly tries to explain a race to the bottom with an ever growing room for laissez faire policies (Babiker et al., 2003: 188). Lobbying efforts by potentially affected industries successfully blocked any attempt of introducing EU-wide carbon taxes. Instead, emission trading schemes were favoured, embracing significantly more degrees of freedom when determining caps and allocation modes (Braun, 2009: 473). The decision of granting nearly the whole emission budget during the first years constitutes the smallest economic burden as compared to any other regulatory instrument (Stavins, 1998: 75), and every more recent change in the framework of the EU ETS was surrounded by further lobbying activities of industrial associations (Smith, 2009: 20-23).

---

[21] Legal basis are the Clean Air Act Amendments of 1990, enacted on 1990-11-15, Public Law PL 101-549, codified at 42 USC 7401 et seq.

While all reasons changed the demand for political solutions, the use of the new instruments can only fully understood by taking into account modified political goals. Here, the most prominent buzz word is economic efficiency entering the argument at the end of the golden age of the nation state (Jordan et al., 2003: 203). This shift in understanding and conceptualising the state only be explained by normative isomorphism (DiMaggio and Powell, 1983: 150). Efficient regulation now stands side by side with traditional types of the normative nomenclature of policymaking, such as public legitimation or legal fit. And exactly in terms of efficiency, a market-based solution is superior to all state- and community-based solutions – being one of the most important rationales for introducing an ETS over other instruments (Hahn and Stavins, 1992: 465-466). This reasoning implies that public bodies retreat from the outcome responsibility in providing for welfare. It is the private actors on the emission market determining the distribution of wealth. The nation state restricts itself to its regulatory (i. e. rule setting) responsibility and its operational responsibility in enforcing the legal framework. Such a rise of market-based solutions constitutes a turning point from the continental European towards the Anglo-Saxon policy model (Bailey, 2007: 537). Consequently, the EU ETS is a misfit not only for corporatist economies like Germany relying on voluntary agreements (Skjærseth and Wettestad, 2008: 282), but also for statist approaches like the French or Italian configurations. Different normative perspectives with respect to which actor should provide for what degree of welfare collide with the concept of self-regulation among different actors (Biermann and Pattberg, 2008: 278).

2.2.4 Concluding Remarks

Section 2.2 provided a review concerning the instruments of environmental regulation from an institutionalist perspective and illustrated the novel character of an ETS. Traditional state-based policy instruments include the determination of performance or technology standards as well as fiscal measures like subsidies or taxes. In doing so, prescribing pollution mitigating technical and economic incentives is always due to the responsibility of the regulator. Because of its complexity, one-to-one compensation schemes largely remain a textbook example of communitarian approaches to environmental regulation. For a long time, market-based mechanisms like ETSs have also only been found in contributions of neoclassical economists. Since they are designed to achieve a predefined pollution target at marginal firm-specific cost, the literature regards a trading scheme as the only efficient environmental regulation. The favourable economic properties can only be achieved by a retreat of the state – its scope is restricted to implementing and enforcing the mechanism, the extent of resulting regulatory costs and the distribution of welfare remain with the market.

In 2003, the EU decided to adopt the first supranational ETS to meet the agreed Kyoto targets by 2012. During the first two trading phases (2005-2007 and 2008-2012), emissions from activities like energy generation, oil refining or metals processing have been regulated by means of a cap-and-trade mechanism. In general, the resulting first phase budget exceeded the aggregate physical emissions, and only the second period is likely to impose a small reduction. The observation of a too generous allocation especially holds for Central and Eastern European countries and for those ETS sectors potentially exposed to carbon leakage. Both findings support the assumption of opportunistic behaviour of the national governments in deriving the emission caps. Many crucial aspects are planned to be changed in the third trading phase running from 2013 until 2020, especially the emission budgets will be determined directly by the Commission, the share of free allocation will constantly decrease, and further activities like the aviation and chemical industries will be added to the regulation.

Finally, this section asked why a regulation of greenhouse gas emissions was assumed by the EU (and not by the nation states) and why a market-based (instead of a state-based) instrument was chosen. Due to the global effect of these emissions, swiftly global lobby groups as well as intergovernmental institutions emerged to promote an international solution. Since the EU had before gained expertise in environmental policymaking, it was regarded as an appropriate actor for the complex task of negotiating and implementing a framework for climate protection. By intention, strict state-based rules were avoided to ensure a small regulatory burden with no significant impact on competitiveness. This goal was accomplished using a grandfathering ETS that additionally satisfies the postulate of an efficient regulation.

2.3 Energy Market Regulation

2.3.1 Economic and Legal Background of European Energy Markets

For environmental regulation, energy markets are the most important object. With respect to the EU ETS, around two thirds of the allowed cap account for the combustion of fossil fuels most commonly employed to generate power. Because of the obvious links between environmental and energy regulation, this section describes the European energy market design and the relevant actors governing this market.

Generally, an energy market is the place for allocating energy resources, and comprises two submarkets: one for fuels and another one for converted energy (Nordhaus, 1973: 530). On the fuels market, primary energy carriers like e. g. coal, gas or petrol are exchanged. Its participants are extractors (mining or drilling firms) supplying these goods either to further processing entities (refineries, generators) or directly to the end-user. The market for converted energy particularly comprises the trade in electricity, heat or steam as provided by utilities.[22] A special feature of both markets is the technical requirement of supplying most forms of energy by means of a pipeline grid. In the political discussion, the term energy market is reduced to power and gas markets (European Commission, 2007). The following analysis focuses on its ETS-related activities, i. e. the electricity downstream submarket (for an analogous usage of this term see only Serrallés, 2006).

Providing power to a customer involves the physical generation in a plant, the transmission over long distances by high-voltage networks, the distribution over short distances by medium- and low-voltage networks, and finally the supply to the end-user (Ferrari and Giulietti, 2005: 248). Electricity cannot be stored and has to be produced just-in-time with demand since deviations lead to an overcharge and – in its worst case – to a breakdown. That is why, otherwise separated grids are interconnected and additional spare capacities are made available (Stalon, 1997: 51-52), and they also allow a short-term and interim

---

22 Other activities of utilities like water supply as well as depletion of wastewater or solid waste are not performed via energy markets since they do not constitute an energy resource in a narrower sense. An exemption is the treatment of sewage as bio-fuel.

fuel switching. To secure the supply, electricity generation relies on plants with an uninterrupted production. In this respect, coal or nuclear stations cover at least the off-peak demand with their base load capacity (Roques et al., 2008: 1833). Further, gas, oil or hydro stations with a reservoir can be swiftly turned on and off to generate additional power in peak-load hours (Söderholm, 2001: 83).

Electricity generation is at the object of several overlapping government interests: most prominently raising energy-related taxes, promoting domestic fuels and ensuring supply security (Helm, 2002: 174). In the traditional European design, the functions generation, transmission, distribution and supply were combined in vertically integrated entities. The reliance on these large enterprises was justified by the natural monopoly assumption: If – due to economies of scale and scope, unit costs proportionately decrease to firm size – the production of a good can be provided at lower cost by one instead of more entities, a natural monopoly exists (Demsetz, 1968: 56). Text book examples are utilities maintaining complex grids as well employing technologies like combined heat and power – giving rise to economies of scope – or nuclear fuels – requiring significant capital intensity and close monitoring (Sioshansi, 2006: 63-64).

This uncompetitive "market" is usually regulated in order to protect the consumers from a misuse of monopoly power by state ownership or state oversight (Jarrell, 1978: 272): state-owned enterprises are either directly owned by a public authority or are steered by directors appointed by a state authority (Shirley, 1999: 115). Due to the diffuse interests of the residual claimant, in this case of the state, the firm tends to operate inefficiently as predicted by the property rights theory (Alchian and Demsetz, 1972: 787). That paved the ground for privately held firms that are more indirectly supervised, first in the US and later also in the UK. Here, private firms are granted the monopolised production of a designated good for a large geographical area – usually for a whole governmental unit or country. Different and sometimes cumulative regulatory instruments are employed to achieve performance. Licensing policies can exclude new entrants, restrict investments or allow only certain activities (Jarrell, 1978: 280). Additionally, a regulation of economic benefits classically restricts the accounting rate of return – for a long time the dominant tool in the US. However, rate of return regulation was criticised for creating distorted economic incentives. Under certain conditions it leads to a biased input factor selection, and hence to higher product costs (Averch and Leland, 1962: 1053). Further, it does not control for economies of scope (Liston, 1993: 28). To overcome these issues, performance-based measures have emerged as a second best regulation of economic benefits for monopolised markets. Here, an efficient production or cost function (a so called frontier) is derived empirically as a benchmark (Jamasb and Pollitt, 2000: 108-109). The policy maker can use these figures in determining a price cap by means of an average allowed revenue – as practiced in the early UK regulation – to reward better performing firms (Bradley and Price, 1988: 99). Long before the rise of the performance-based regulation theory, politically determined price caps were used in Germany as early as 1935 since energy utilities were obliged to submit selling prices to an ex ante review (EnWiG:[23] § 7; Metzger, 2008: 461).

Due to the national scope of energy politics, the electricity markets in Europe remained isolated in physical and economic aspects. Authorised by the provisions of the Single European Act[24] (Art. 13), the European Commission advanced an internal market for energy (Newbery, 2001: 85). With Directive 1996/92/EC,[25] the unbundling of vertically integrated utilities, third party access to transmission and supply grids, as well as the free choice of power suppliers for large customers were prescribed (Art. 14, 17-19; Serrallés, 2006: 2545). It was explicitly left to the Member States' discretion to decide on the domestic market design and whether they wanted to establish a fully competitive arrangement which would also include small end-users. Somewhat unexpectedly, the emerging liberalisation of the electricity markets facilitated a further concentration, and Table 7 outlines this wave of mergers and acquisitions by showing its most prominent examples. Some of these transactions have been fostered by strategically acting national governments to create national or European "champions" (Verde, 2008: 1126), like the Suez/GDF deal in France.

---

23 Energy Economics Act (Gesetz zur Förderung der Energiewirtschaft – EnWiG), RGBl I (1935), 1451-1456.

24 Single European Act, OJ L 169: 1-29.

25 Directive 1996/92/EC of the European Parliament and of the Council of 19 December 1996 concerning common rules for the internal market in electricity, OJ L 027: 20-29.

Table 7: Major Mergers and Acquisitions in the European Electricity Industry since 1996

This table presents the major changes in the European electricity industry since the first liberalisation directive of 1996. Viewed from the perspective of the acquiring firm (except for mergers), either national or cross-border transactions are listed in chronological order. The following common firm name abbreviations are used: AEM (Azienda Elettrica Municipale), ASM Brescia (Azienda dei Servizi Municipalizzati di Brescia), DONG (Dansk Olie og Naturgas), ÉDF (Électricité de France), EnBW (Energie Baden-Württemberg), ENEL (Ente Nazionale per l'Energia Elettrica), GDF (Gaz de France), HEW (Hamburgische Electricitäts-Werke), VEAG (Vereinigte Energiewerke AG), VEBA (Vereinigte Elektrizitäts- und Bergwerks AG), VIAG (Vereinigte Industrieunternehmungen AG) and VWE (Vereinigte Elektrizitätswerke Westfalen).

Country	Buying firm	National merger	Cross-border merger
Denmark	DONG	2006: Elsam, Energi E2	
Germany	EnBW	1997: Badenwerk, Energiev. Schwaben	
	E.ON	2000: VEBA, VIAG	2002: Powergen (UK)
	RWE	2000: VWE	2000: Innogy (UK) 2009: Essent (Netherlands)
France	ÉDF		2000: EnBW (Germany) 2005: Edison (Italy; jointly with A2A, Italy) 2009: British Energy (UK)
	Suez	2008: GDF	2005: Electrabel (Belgium)
Italy	A2A	2007: AEM, ASM Brescia	
	ENEL		2007: Endesa (Spain)
Netherlands	Essent	1999: PNEM, EDON	
Spain	Iberdrola		2007: Scottish Power (UK)
	Gas Natural	2009: Unión Fenosa	
Sweden	Vattenfall		2002: HEW (Germany) 2002: VEAG (Germany) 2003: Bewag (Germany) 2009: Nuon (Netherlands)
Un. Kingdom	Scottish and Southern En.	1998: Scottish Hydro-El., Southern El.	

The Commission became aware that a functioning competitive market is at risk if monopolistic structures spread out to only weakly interconnected markets (Domanico, 2007: 5069), and brought about the second electricity market directive in 2003. Four central features have to be implemented by the Member States (see also Jamasb and Pollitt, 2005: 7):

- A legal unbundling of the electricity industry is prescribed into generating entities, transmission system operators (TSOs) and distribution system operators (DSOs – Directive 2003/54/EC:[26] Art. 10, 15, 18; van Koten and Ortmann, 2008: 3129). Transmitters operate on wholesale markets to (economically) settle and (physically) balance the power load, and distributors interact between TSOs on the one hand and the final customer on the other hand. Instead of relying on long-term contracts, producers are now to offer their capacity on national or regional wholesale power exchanges to provide for the most efficient price for suppliers and large consumers (Meeus et al., 2005: 28-29).
- All types of end-users, including households, must be granted access to a competitive supply market from 2007-07 onwards (Art. 21(1)(c)).
- Any entity apart from the incumbent monopolist may be authorised to install new generation capacities and has access to the existing grid under a determined tariff scheme (Art. 6, 20).
- To foster and monitor competition, National Regulatory Authorities (NRAs) are to be set up (Art. 23). Generally, NRAs can be disposed of the competence to set network costs and tariffs, to oversee the status of unbundling and market dominance, to enact own rules and regulations, as well as to fine firms violating their licence or the security of supply (CEER, 2005). However, the institutional design varies among the Member States: In some countries like in the UK, several functions were concentrated in the Office of Gas and Electricity Markets (Ofgem). In others,

---

[26] Directive 2003/54/EC of the European Parliament and of the Council of 26 June 2003 concerning common rules for the internal market in electricity and repealing Directive 96/92/EC, OJ L 176: 37-55. This directive was accompanied by Regulation (EC) No. 1228/2003 of the European Parliament and of the Council of 26 June 2003 on conditions for access to the network for cross-border exchanges in electricity, OJ L 176: 1-10.

already existing organisations like competition authorities assumed some of these responsibilities.

Concerns over lacking competition and harmonisation still persist after Directive 2003/54/EC has taken full effect. In most Member States only few small consumers are currently willing or able to switch their suppliers (European Commission, 2008: 5-6). To address country effects in regulation, the scope and direction of NRA activities is coordinated by two EU-level bodies: the Council of European Energy Regulators (CEER) forms the association of national authorities and discusses and prepares emerging issues for the European Regulators' Group for Electricity and Gas (ERGEG) as the Commission's advisory group for energy markets. Both committees currently favour an ex post regulatory approach, and ex ante measures like retail market price controls are regarded as distorting free markets (ERGEG, 2007: 6). In 2011, the Agency for the Cooperation of Energy Regulators (ACER) will assume its service and replace the existing regulatory framework at the EU level (Hancher and de Hauteclocque, 2010: 3). The agency is to support the Commission in implementing a single energy market and to coordinate the policies of the constituent NRAs that maintain their exclusive enforcement competences (Regulation (EC) No. 713/2009:[27] Art. 6, 7).

2.3.2 Reasons for Transformations in Energy Market Governance

At least since the end of World War II, European monopolised and integrated power firms acted on behalf of state agencies. The absence of market arrangements in power supply appeared to be set in stone up to the end of the 1980s. However, several causes promoted a shift in the regulatory regime[28] throughout

[27] Regulation (EC) No. 713/2009 of the European Parliament and of the Council of 13 July 2009 establishing an Agency for the Cooperation of Energy Regulators, OJ L 211: 1-14.

[28] This analysis does not focus on the changes in ownership structures of utilities as e. g. presented in Ehni et al. (2004: 135), but instead on the policy configuration employed. In this respect, an early privatised utility like the Belgian Electrabel still operates in an arrangement with little to no market elements that is dominated by the public rule maker. As opposed to that, in Germany public utility ownership only faded slowly and gradually, but the electricity market has been traditionally rooted in corporatism and self-regulation (Humphreys and Padgett, 2006: 398).

the continent. An institutional cause is the "1992" process of creating a single European market as agreed on in the SEA. Its central goals were achieving the free movement of persons and goods as well as the abolishment of internal frontiers (Sun and Pelkmans, 1995: 70). Consequently, an unrestricted and non-discriminatory transmission of electricity could be derived from the treaty's provisions. The Commission did not only promote upgrading and establishing interconnections between the national energy markets, but further proposed the restructuring of the whole industry (Jabko, 2006: 102). Despite its weak negotiation position, the Commission successfully lobbied for a complete liberalisation, and was able to highlight the efficiency gains of full privatisations (Häge and Schneider, 2004: 282).

An intellectual cause in transforming the energy market regimes is the dominance of marked-based thinking and is hence closely connected to the liberal ideals promoted by the Commission. This factor does not rely on the institutional framework of the EU, but instead directly takes effect as policy learning from other countries (Humphreys and Padgett, 2006: 385). In this respect, the forerunner UK unilaterally began privatising its state-operated monopoly provider in 1989. During this process, the functions of the integrated Central Electricity Generating Board (CEGB, England and Wales) were unbundled and wholesale and consumer markets were created (Woo et al., 2003: 1105; Amundsen and Bergman, 2006: 148). The market concept was favoured since a large scale of directly interconnected producers promised a balancing of national power generating over- and under-capacities at less cost. Additionally, multi utilities could emerge – simultaneously supplying power, gas, water and processing waste – that were eager to expand their activities abroad.

Generation technology issues exhibit a material cause. Technologies like the combined cycle gas turbine (CCGT) were invented in the 1970s and attained marketability in the early 1990s (Colpier and Cornland, 2002: 309-310). Others like combined heat and power (CHP) and wind farms witnessed a revival around the same time (Weber and Hoogma, 1998: 553). Additionally, the Černobyl' disaster of 1986 raised the question of nuclear power plant safety in several countries (Bunyard, 1990). All these developments favoured small and decentralised plants, and neutralised the natural monopoly argument.

As a financial cause, global competition among energy consuming industries can be named. Due to the physical restrictions of transmission and supply grids, the price for electricity is not directly subject to globalisation, but exporting industry firms suffer from steep price differentials between countries or regions. In this respect chemical, aluminium and glass processing are among the sectors most severely exposed to electricity prices (Grubb and Neuhoff, 2006: 10). That is why the German government, protecting its significant industries, supported an energy market liberalisation assuming a drop in the high domestic power prices (Eberlein, 2000: 85).

These causes affect nation states differently, depending on their predominant regime of providing welfare. To this end, market-, community- and state-based configurations are expected to react differently to the deregulation of energy markets, and this process is first of all a transition in leading actors. During the golden age of the nation state, power was generated and supplied either by monopolists in public ownership or by private regional monopolies under state oversight. Prominent examples for a strong state involvement in energy markets are the former state-owned enterprises ÉDF, ENEL or the CEGB. Differently, corporatist countries like e. g. Germany relied on the community-based approach balancing the organised interests of an industry with the administration (Bartle, 2002: 7). These firms were gradually privatised starting in the 1960s, developing a genuine corporatist energy market configuration. Following Teulings and Hartog (1998: 31) and Bruno and Sachs (1985: 241), the Nordic countries, Austria, Germany and the Netherlands are classified as corporatist countries; all other energy markets were state-based throughout the golden age. Terming a country as corporatist reflects the effective ownership structures and geographical scopes of the national energy industries almost perfectly (Serrallés, 2006: 2543).

With the early 1990s, first liberalisation efforts changed this initial setting of formerly state- or community-based arrangements. As depicted in Table 8, this transition has taken place at different speeds, resulting in five different country clusters. The early policy shift of the forerunner UK proved to be model for all subsequent reforms in energy market policy. Shortly after that, in 1991, the non-EU member Norway followed with a similar step towards a market-based regu-

Table 8: Steps in Energy Market Transformation

This table presents the shift in European energy market regimes. For the time up to 1988, it follows the classifications of Teulings and Hartog (1998: 31) and Bruno and Sachs (1985: 241).

Country	Golden Age	Liberalisation	
	Until 1988	Early stage 1989-2003	Full effect starting 2003
Forerunner			
United Kingdom	State	Market	Market
Nordic market			
Denmark	Community	Market	Single Nordic market
Finland	Community	Market	
Norway	Community	Market	
Sweden	Community	Market	
Decreasing corporatist influence			
Austria	Community	Community	Market
Germany	Community	Community	Market
Netherlands	Community	Community	Market
Decreasing state influence			
Czech Republic	State	State	Market
Italy	State	State	Market
Poland	State	State	Market
Portugal	State	State	Market
Spain	State	State	Market
Strong state influence			
Belgium	State	State	State
France	State	State	State
Greece	State	State	State
Ireland	State	State	State

lation. The other Nordic countries did not only transpose Directive 1996/92/EC but created a framework of broad customer access by far exceeding the European provisions (von der Fehr et al., 2005: 1). By this means, a single Nordic electricity market emerged among these four countries with the liquid spot and derivatives exchange NordPool. Presently, along with the British one, both market-based regimes witness the least number of still state-owned enterprises, and both are comparatively competitive (Amundsen and Bergman, 2006: 155).

All other Member States more or less maintained their traditional either community- or state-based electricity market regimes at the early stage. Even though the earliest electricity market directive of 1996 had mandated a liberalisation of power supply, the underlying core institution – i. e. a functioning market – did not come into effect. Only with the transposition of the subsequent electricity market directive of 2003, most Member States established market-based configurations. This applies to the former corporatist as well as to a number of former state-based energy markets. A major retreat for public monopolies did not only take place in Mediterranean countries like Italy (Ferrari and Giulietti, 2005: 251-252), but also in transition economies like the Czech Republic or Poland. The fact that power markets have been established does not imply that they work efficiently in an economic sense, and some of them, like the Portuguese, are presently regarded as highly oligopolised (for a comprehensive review see European Commission, 2007). However, all of the market-based regimes rely on mostly fully privatised generators, unbundled transmitters and suppliers, and little regulatory intervention. The last group of Member States did transform both directives to national legislation, but state-based energy market regimes prevail here, comprising Belgium, France, Greece and Ireland. In these countries, fully publicly owned or only partly privatised, but still monopolised firms render market arrangements superfluous.

Two central observations can be made. Both energy liberalisation directives coerced by the European Union meant a transposition of the British (or more specific: of the English and Welsh) regulation to all other Member States (Newbery, 2001: 91). Some of them, like the Nordic countries, unilaterally adopted similar rules in the phase of transition, but all other governments were more or less reluctant in governance mode changes. The European energy market marks a retreat for corporatist, but also for most state-based configurations. Additionally, with the emergence of market-based social orders, the former actor's community and state lost both their operational responsibility of generating and supplying electricity and their outcome responsibility of providing welfare to the market. Still within the nation state's scope rests the regulatory responsibility. Concurrently to liberalisation, NRAs were set up to provide in the first place for a secure supply, and above that for an economically functioning market. Here, the incentive policies as preferred by the Commission also represent a mimicking of the UK and the Nordic countries regulatory approaches (Larsen et al., 2006: 2865).

The shift in environmental policy and in energy regulation did not only coincide with one another in a temporal manner, but also with respect to the chosen instruments. Even though the energy markets are still more fragmented and less harmonised than the one for allowances, the direction towards a market-based configuration with as little state interventions possible is the same.

2.4 Are the New Environmental and Energy Policies Cases of Regulatory Capture?

Even an ex ante efficient framework coerced by the sate can prove as ex post ineffective if the rule maker is unable to design an unbiased policy instrument. The theory of regulation offers two mutually contradicting hypotheses as to possible outcomes of political projects: the one posits that the regulator intends and succeeds in acting in the public interest (public interest theory). The other one claims that the politicians' goals may have been captured by some interest groups, and hence enact a regulation that is to a certain degree ineffective, at least from the public interest point of view (capture theory).

The public interest theory postulates that any action by government has to provide for the welfare of the common. Deeply rooted in philosophy and legal sciences as a normative concept of how the state should act (Hantke-Domas, 2003: 166), it entered economic theory by means of the so called Virginia school of thought. According to its contributions, given the free choice of voters, a regulatory decision-making process ends in a majority equilibrium since unanimity cannot be achieved (Buchanan, 1954: 121). Some authors link the public interest theory with the (Pigouvian) economics of welfare (Joskow and Noll, 1981: 36; Barr, 1992: 756): The state assumes an action if welfare is distributed in an inefficient or unequal manner due to market failure, like in the case of externalities, natural monopolies or illiquid markets. Consequently, the public interest is met if the discharger fully internalises his external effect, the respective market regains its liquidity, and so on. Virginian authors oppose this assertion and regard the rule maker as coordinating transfers and mitigating illegimate rents (Mitchell, 1988: 106; Mitchell and Munger, 1991: 523). However, this theory neither evolved as a deeply-rooted economic concept nor did it motivate quantitative or empirical research.

The capture or interest group theory hypothesises that the public actor embarks on the individual interests of one or several groups. Three threads of literature can be found that have remained to date largely unconnected: Firstly, in line with classical Marxist theory, the capitalist class does not only dispose of the ownership over resources, but is also able to influence the state to its needs in obtaining a maximum rent (Pitelis, 1994: 88-89; see for a critical view Posner, 1974: 341). However, this approach allows neither for the emergence of institutions (Cullenberg, 1999) nor for the development of testable research hypotheses. Secondly, the political theory of group formation evolved along two contrary contributions (Walker, 1983: 390-391). Truman (1967: 36) assumes an unplanned creation of pressure groups as soon as several individuals are discontent with a certain aspect of their economic or political situation, and hence predicts large interest groups. Olson (1980: 28, 143) hypothesises that the smaller a pressure group, the more efficiently it works. At its core, the political strand of literature focuses on the causes for the rise and fall of lobbying groups, but does not allow for a thorough analysis as to the interactions between the policy maker and the distinct interests. Finally, the economic capture theory emerging in the 1970s employs a positivist approach (for comprehensive reviews on this theory see James, 2000, and Dal Bó, 2006). It can be drawn back to the so called Chicago school of thought, with contributions by Stigler (1971), Posner (1974), Peltzman (1976) and Becker (1983). Throughout the early stages of this theory, a pressure group was most commonly defined as either a monopolist or a competitive industry; other lobbying parties were disregarded. Basically, it assumes that a policy maker can distribute economic rents by imposing a burden on one party or granting benefits to another one by means of subsidies, tariffs, taxes or entry regulation (Stigler, 1971: 5-6). Which action is chosen by the regulator is either due to its bounded rationality or due to incentives of lobbying parties, i. e. bribes, votes, revolving door opportunities or threats – the political or regulatory rent. This policy outcome is traded on a market for regulation: An interest group seeks for an economic rent as long as its benefit is higher than the resulting cost of coordination and of free riders. Consistently, the policy maker supplies the good regulation to the one group providing the highest regulatory rent netted by the political costs of maintaining the legislation (Peltzman, 1976: 212). A suchlike competition for favourable or protective legislation shows all fundamental traits of economic market mechanisms, and consequently the one group exerting pressure in the most efficient way receives the most comprehensive grant (Becker, 1983: 380).

Early models in positive economic capture theory suffer from simplifications and strong assumptions. Both, regulators and pressure groups are treated as black boxes. The former ones are modelled as a monolithic apparatus without distinct properties, and the lobbying groups are exclusively made up by corporations, neglecting e. g. the firms' victims. Further, the employed equilibrium models are unable to control for agency settings. To overcome these issues, two categories of neoinstitutionalist models emerged promising a broader insight to economic realities (Bennedsen and Feldmann, 2002: 920). Papers in the first category ask the question how a governmental output as envisaged by an interest group can be achieved. Two-tier models employ one agent – the lobbying producer – and one principal – the legislator. Since the regulator is unaware of the firm's cost structure, the public body has to rely on reports prepared by the entity, resulting in an inefficient rule set (see Baron and Myerson, 1982: 922, for a regulated monopoly price higher than firm-specific marginal costs). Three-tier models introduce a regulatory authority acting on behalf of the legislator, adding further distortions to the political process (see only Spiller, 1990: 75). Here, the authority exhibits moral hazard – an outcome in the principal's sense is claimed as own achievement, a regulatory failure has to be borne by the regulated agents (Tirole, 1986: 206-207). The agents' lobbying at the authority is modelled to be successful in realising a weak implementation of the legislator's regulation (Laffont and Tirole, 1991: 1104). Other authors focus on how pressure groups safeguard that a desired outcome is eventually reached (e. g. Scholz, 1991: 120, 128) Even though agents reject the classical enforcement due to control problems, they embark on a cooperative approach providing for an efficient model solution. Articles in the second category ask for an interest group strategy considering the legislator's majority stakes. In this respect, lobbyists offer larger benefits to those parliamentarians being opposed than to those being inclined (Snyder, 1991: 102). If the representatives are unlikely to change their voting behaviour, the lobbyist is predicted to instead promote his concepts in other influential institutions along the political process (Bennedsen and Feldmann, 2002: 936-937).

Having briefly outlined the public interest and capture theories, it can be asked whether the EU ETS and the European energy market can be aligned with either of the concepts of regulation. In economic terms, consumers and environmental protection groups pursue the identical goal of coercing the producers to internalise their pollution-induced external effects. Consequently, the mere existence of an environmental regulation can be ex ante termed as achievement of the public choice implementing an instrument to realise the majority's intention. As soon as consumers successfully lobby for production-related pollution abatement (see for an agency model Laffont and Tirole, 1991: 1110), it is unclear whether they act on behalf of the aggregate welfare or instead try to overcompensate only a small fraction of the population. Still, the first phase allowance allocation is largely regarded as regulatory capture induced by the industry's interests: a fully free supply can give rise to opportunity costs of final products, and the over-generous supply secures cash receipts disposing supernumerous rights. Similarly, the carbon leakage argument entered the third phase regulation documents (consolidated post-Kyoto Directive 2003/87/EC, Art. 10a(6) and (15)) and further supports the capture by industry interest. If this assumption holds, a double subsidy like this constitutes a text book example for successful pressure group lobbying.

The regulation of natural monopolies in energy markets has received much attention in the economic literature. From the public interest theory point of view, a monopolised utility is potentially able to achieve economies of scale and scope, but it is unclear whether this advantageous cost structure is reflected in selling prices considering the market power of the single seller. To this end, an efficient neoclassical regulation is set out to achieve the same price level as the (unobserved) one of competitive markets. If the monopolist acts as an interest group, it is predicted to generate rents by influencing the political process to set an above-competition price. Models in line with this assumption consequently yield a regulated price in between of the competitive and the monopoly levels (Peltzman, 1976: 223-224; Baron and Myerson, 1982: 922). There are first indications that the present shift towards a market-based energy market regulation is a case of regulatory capture: First, there is the wave of mergers and acquisitions, which further strengthened the largest entities. Second, the complete absence of new entrants needs to be noted – neither there are large-scale power imports nor there is additional capacity installed. Both factors result in stable market shares of the incumbents.

2.5 Some Resulting Questions for Empirical Research

This chapter gave an overview of the available and currently preferred modes for environmental and energy regulation in terms of their economic, legal and political implications. Pollution control in the EU is at its crossroads. Traditional and established policies are being eclipsed by the emerging ETS. From an economic viewpoint, the marketability of emissions by means of certificates is more efficient than any other instrument, since the abatement takes place at marginal costs, and the regulator's only tasks are to define the cap and to enforce the surrender. However, the implementation to European norms gives rise to questions of regulatory capture, considering that many Member States endowed firms with more rights than ex post demanded for, and, further, the stake of costly allocation is still small. Finally, a trading scheme as a market-based solution is novel to the traditional continental European understanding of the nation state as provider of welfare. A similar transformation of policy modes has taken place in the EU electricity regulation. In this sector, the nation states shift from highly regulated, either state-owned or corporatist arrangements, towards a liberalised market-based mode. Consequently, the European politics embarked on the idea of free markets in the early 1990s and dropped the natural monopoly argument. Instead of adopting a mere mutual market opening, a far-reaching conversion of the European utilities was agreed on.

Despite the broad theoretic discussion, a gap in literature exists with respect to the *empirical* analysis of the economic consequences induced by both regulations. Accordingly it can be asked if the EU ETS introduces distributional effects between the regulated energy firms and other market participants. Further it can be addressed if energy market regimes determine the extent of distributional effects introduced by an ETS.

Quantifying the extent and direction of distributional effects answers whether beneficial or detrimental effects occur at the firm level. An analysis of economic consequences has implications for European and national regulators, and shows whether the desired result has been achieved. Sections 4.5 and 4.6 will address these questions separately by employing multifactor capital market models.

The novel scheme has also created facts that need to be reflected in the management and assessment of business entities. The following Chapter 3 will therefore deal with the accounting implications of emission trading schemes. To this end, an extension of accrual and imputed cost accounting systems to allocate emission costs is developed in its first half. Throughout the remainder of this chapter, major financial accounting treatments are presented in order to prepare the third question for empirical research – addressing the usefulness of respective accounting treatment information conveyed in capital market valuation.

3 Mapping Emission Rights to Accounting

3.1 Framework for Accounting Analysis

Accounting for emission rights poses new challenges as it differs from treatments for other environmental policies: an ETS develops its coercive effect via the restricted budget, and fluctuations in prices are derived from the quantities traded. For the time being, accounting relies on the monetarisation of observables, and hence can only capture the composed price-quantity realisation of allocations, trades and obligations to surrender. The way in which the ETS framework is transposed to accounting treatments depends on two factors: the purposes of accounting and the degree of codification and standardisation. Different accounting purposes create different information needs, and hence promote different recognition and measurement techniques (Schneider, 1994: 28). The stricter an accounting system is codified or the higher it is standardised, the accountant has fewer degrees of freedom in formulating specific techniques. With respect to these two criteria, management accounting[29] has to be distinguished from financial accounting. That is why in the remainder, management and financial accounting are analysed separately.

Accounting for internal uses most commonly follows either of two distinct approaches: accrual costs or imputed costs. Their application patterns vary with national traditions and resulting path dependencies, and hence neither of them can be excluded in a comprehensive review of methods applicable for European firms. By means of a normative analysis, the available techniques are at first applied to the case of emission rights, and are subsequently evaluated with respect to the performance reporting functions of stewardship and decision making.

Financial accounting, for its part, is strongly regulated in the EU and is based on the International Financial Reporting Standards (IFRS) developed by the private standard setter International Financial Accounting Board (IASB). All listed firms are required to adopt these international rule set for group accounts since the fiscal year 2005 (Regulation (EC) No. 1606/2002,[30] Art. 4). In this re-

---

[29] Management accounting, cost accounting, and performance reporting are used as synonyms in the remainder.

[30] Regulation (EC) No. 1606/2002 of the European Parliament and of the Council of 19 July 2002 on the application of international accounting standards, OJ L 243: 1-4.

spect, all (formal) traditions for group accounts came to a halt across the Member States at that date. The subsequent normative and empirical analysis of financial accounting focuses on the different properties of presently available treatments that can be used alternatively in IFRS-conform statements.

3.2 Consequences for Performance Reporting

3.2.1 Functions of Management Accounting

The question of how to map emission rights to management accounting has been disregarded for a long time. Even though several cap-and-trade schemes have been in force in the US, and the EU has launched its own system by 2005, still no consistent treatment of allowances in performance reports can be found in the literature. This emerging issue is on the accounting agenda of many entities in the industrial sector. Additionally, the scope of the existing EU ETS is about to be even enlarged in the future. Assessing management accounting techniques can either be done in a normative or a positive way. The former approach subjectively values the "suitability" of a dedicated method, and is in this regard prescriptive (Watts and Zimmerman, 1979: 273). Conversely, a positive evaluation relies on a description of incentives and interests of both accounting users and preparers. The remainder of this section pursues a positive approach with respect to the functions of performance reporting.

Traditionally, cost accounting is used for recording, budgeting, stewardship and decision making purposes. Controlling the performance of managers and deciding among various alternatives under uncertainty are two management accounting purposes that stand in a conflicting position and are most thoroughly discussed literature (Kaplan, 1984: 401-404; Zimmerman, 2001: 424). In today's accounting systems, recording and budgeting are more or less only derived from the stewardship and decision making purposes. Documenting the transformation process was central to the development of early industrialised firms (Boyns and Edwards, 1997: 13), and budgeting as the short- or medium-term planning of costs and capital is an essential feature in the measurement and review steps in the control cycle (Hesford et al., 2006: 7). However, both recording and budgeting are constructed and biased to the extent that management accounting serves for stewardship and decision making purposes. Hence, incurring and measuring emission cost approaches must also be evaluated against the background of management control and decision making.

Some authors in management accounting theory derive reporting instruments that create equal incentives for principals and agents in a stewardship setting (Reichelstein, 1997: 157). Others shed light on measurement techniques that do not in the first place reflect valuation purposes but are dedicated to inform on underlying events (Mattessich, 2006: 222). In the following, quality criteria of management accounting systems are developed as benchmark for evaluating different techniques of recognition and measurement. They rely on basic characteristics of performance measures as well as on features developed by both schools of thought, and some of them stand in a conflicting position and cannot theoretically be fulfilled at once. Two fundamental postulates are mentioned in both stewardship and information economics settings:

1. Accounting information is perceived as reliable as long as the economic agents to be evaluated are unable to manipulate the reports (Laux, 2006: 90). For example, this assumption materialised in financial accounting standards, like the faithful representation of economic events in accounting, freedom from bias and completeness characteristics advanced by the IASB Framework (FW – paras. 31-38).[31] From a theoretical point of view, this postulate is at risk if the management is able to choose over accounting alternatives, resulting either in income smoothing or in income maximising (Beidleman, 1973: 653-654; Watts and Zimmerman, 1978: 114; Merchant, 1990: 300).
2. Comparability extends to both an inter-temporal and a material dimension. Consistency and continuity in measurement techniques means that all equal economic events have to be mapped to performance reporting in the same way (Sunder, 1973: 2-3; Laux, 2006: 90). Otherwise those controlling managerial action or making decisions have to dispose of a full knowledge concerning the measurement inconsistencies.

The further postulates and observations differ between the stewardship and the decision making perspective. Performance reporting instruments for stew-

[31] The fourth quality characteristic pertaining to reliability, prudence (IASB FW.37), is explicitly excluded in this context. It is derived from the theoretical postulate of value creation described below.

ardship purposes aim at resolving conflicts when ownership and management is are dispersed. It is hypothesised that a principal (the owner) entrusts an agent with increasing a firm's income by his actions, but suffers for his part from asymmetrically distributed information in this setting: Since the agent's endeavours are unobservable to the principal, the agent can undertake any action in maximising his own wealth opportunistically (Alchian and Demsetz, 1972: 787-788). Conversely, the principal is a profit maximiser, too, aims at monitoring the agent as closely as possible, and tries to resolve the information asymmetry (Akerlof, 1970: 490-492). In this respect, management accounting plays a decisive role in distinguishing costs and benefits either arising from management performance or from exogenous shocks. Laux (2006: 89-90) describes characteristics of performance measures towards their fit for contracting and stewardship uses:

1. The first stewardship criterion relates to the clean surplus postulate: as long as discounted accounting earnings and discounted cash flows are equal over a firm's total lifetime, both measures are equally suitable in evaluating investments based on either present value (Preinreich, 1938: 240; Lücke, 1955). An accounting system works as "clean surplus" given all (non-owner related) changes to equity are booked through profit or loss (Feltham and Ohlson, 1995: 694).
2. The hypothesis of value creation holds if a decision's economic consequence is immediately and appropriately reflected in accounting income. In the sense of shareholder value, a symmetric timeliness of accounting is desirable to evaluate management performance as early as possible (Rappaport, 1981: 140-141). As opposed to the Basu (1997: 6-7) sense of conditional conservatism, principals desire assessing the quality of past events in maximising their wealth without a bias for positive events.
3. Pareto-efficiency in sharing risk between managers and owners originates from different utility functions inherent in both parties. Earnings-based management compensation leads to a risk-adverse behaviour of the managing agent, and they aim at smoothing accounting income in order to establish a stable remuneration – which violates the postulate of reliable accounting information (Holmström, 1979: 86).

Decision making in the management accounting context generally relates to predicting the outcome of alternatives, like different processes, jobs or products, as well as the identification of determinants driving the results (Kirchner, 1961: 47). All these approaches assume a risk-averse decision maker with time preference (Samuelson, 1937: 156; Feltham, 1968: 692). By means of reporting, the decision maker aims at resolving his uncertainties. In this respect, an economic agent is ex ante not fully informed about the significance of future events for the current period, for example the volatility and distribution of cash inflows arising from multi-period investments. Accounting procedures are unable to perfectly reveal the underlying interdependencies. Additionally, managers and owners jointly act as principals with respect to the heads of cost centres or divisions. The information content argument relies on the fineness of information as set out in the theorem of Blackwell (1953): disregarding information acquisition costs, a decision maker prefers an information source over another that maps a transaction in the most detailed way:

1. Key assumption of information economics is that a figure not necessarily reports about the value of a transaction in an unbiased, but instead in a timely, and with respect to its sign, correct way (Demski and Sappington, 1990: 369). If event and information release are close to each other, the uncertainty about present and past events at the time of making a decision is small, otherwise the decision is delayed until accounting finally reports about the event. Thus, the earliest possible revelation of temporary averse shocks is an efficient characteristic of an accounting system (Baiman and Evans, 1983: 384).

2. Freedom from noise is a necessary condition in evaluating the relative information content of an accounting treatment with respect to another alternative (Marschak, 1971: 201). Intuitively, the economic decisions are based on the outcome of such a report, and possibly false signals reduce efficiency since the decision maker controls for this possibility in increasing his certainty equivalents (Sandmo, 1971: 72; Leland, 1972: 289-290). From a technical point of view, noise can have two reasons: an inadequate measurement causing semantic noise, or errors due to communication problems, the technical noise (Shannon and Weaver, 1998: 26). These er-

rors can be corrected by means of redundant information (Nakano, 1972: 698, 706).

The remainder of Section 3.1 analyses two different management accounting approaches: accrual and imputed costs. Under the former paradigm, cost incurrence is strictly based on underlying cash flows, whereas the latter one allows additional cost components that are derived from the consumption of resources. In the first instance, both performance reporting methods are depicted with respect to their recognition and measurement techniques. After that, these procedures are evaluated with respect to their suitability with the seven mentioned criteria.

3.2.2 Emission Rights as Accrual Costs

3.2.2.1 Cost Incurrence

Commonly, costs are defined as cash outlay to obtain input goods, whereas all other transactions not conforming to this requirement do not enter managerial accounts (e. g. Noble, 1926: 52; Koch, 1958: 361; Bhimani et al., 2008: 38). Even though some authors discuss the additional inclusion of opportunity and sunk costs (McWatters et al., 2008: 42-46), these two cost definitions are irrelevant under the traditional paradigm of accrual costs. Applying this recognition criterion to the case of an emission trading scheme, it is evident that only those allowances incur costs that have been or must be purchased during the financial year. Hence, certificates sold or auctioned off during the regulatory allocation process as well as those offered at an exchange intuitively qualify as costs.

Due to a predominantly free endowment of allowances during the first two trading phases, the vast majority of transactions regulated by the EU ETS remains unrecorded irrespective of the reduction target. Currently management accounting is likely to report only a small impact of emission trading in most cases. Considering the given regulatory framework, management accounting that employs accrual costs for emission rights informs on a firm's short positions; the system only conveys a cost figure if it was necessary or whether it is very likely to purchase additional rights. At first sight, a cost incidence might reveal the management's failure in adequately adapting to the regulatory obligations in reducing pollution. However, it is not straight forward what a "low" free endowment of emission rights stands for. It can be firstly driven by ex ante political considerations towards the free fraction in allocation; secondly, fluctuations in demand drive output and hence emissions (Ellerman and Buchner, 2008: 270-271). The last, and from an accounting perspective the most interesting point, is the question of pollution abatement at firm level as the only truly endogenous variable. This latter aspect of the cost incidence captures the degree to which management was able to adapt to the prescribed reduction target.

Since performance reporting is not bound to the strict valuation rules of financial accounting, it can be decided on the way of measuring those transactions in allowances connected to cash outflows. Apart from historical costs that are strictly based on past acquisitions, fair values or replacement costs can be equally applied (Holzer and Norreklit, 1991: 7). Assigning emission costs at fair values can be implemented either by incorporating the current market price as observed at one of the designated exchanges, or by relying on a weighted average of historical market prices of the annual trading cycle. Similarly, the replacement value of emission rights can be determined with little discretion considering the quotations for futures. Especially year-ahead contracts portray the market's estimate towards the medium-term development of certificates prices.

Valuation has to fit with the given cost incurrence scheme as well as with the function of management accounting, here stewardship or decision making. With regard to the former function, the evaluation of a manager's action is accomplished by analysing the variance between realised and planned values. An analysis like this has to provide for the fact that exogenous components not attributable to these actions are eliminated to (better) observe the agent's behaviour (Ewert and Wagenhofer, 2000: 430). Generally, prices play a subordinate role in a setting that focuses on variances in in- or output volumes (Kilger et al., 2002: 135-138). Accrual costs restricted to inform on under-allocation should determine the short position with respect to the occurred or likely cash outflow: that is by means of historical costs for already settled transactions, and by means of current or replacement costs for those allowances voided in the course of production that still have to be purchased during this trading cycle to settle the obligation. With respect to decision making, two important objectives are the product mix and pricing. Hence, this accounting function draws on forward-looking measurement techniques for short- and medium-term assessments (Dye, 2004:

1627). Here, fair values or replacement costs reveal the predictions of all participants in the ETS better than historical costs.

Connected to the measurement of emission allowances is the issue of derecognition due to sale. Given the fact that the management was able – for whatever reason – to use fewer emission rights than presently in stock, they can be disposed off to other emission market participants. It can be questioned whether recognising the difference between purchase and exit price as revenue is appropriate. Firstly, if certificates are traded for speculative reasons, these gains and losses do not relate to the provision of goods and services. Assigning costs and revenues by means of cost centres or cost drivers, most cost accounting systems more or less implicitly only focus on the production process (Luft and Shields, 2002: 65-66). In this respect, speculative net profits due to an ETS cannot give rise to revenues in management accounting.

Emission revenues should not be completely excluded, though: in case of an entity is over-endowed, the long position is most likely to be sold off. These indirect revenues can be assigned to cost objects for they arise in alignment with production. In line with revenues accrued in the course of the classical production cycle, revenues due to the by-product "excessive emission allowances" are realised as soon as the respective transaction at an exchange has taken place (Schweitzer and Küpper, 1989: 100). To provide for this possibility, an accrual cost accounting still has to record freely allocated emission rights (that is recognised at zero cost), for they carry a value that is only revealed when not surrendered to a regulatory authority.

3.2.2.2 Accrual Costs in Stewardship and Decision Making Settings

The accrual cost accounting procedures apply a restrictive selection criterion in mapping emission rights to reporting. Only those allowances surrendered incur costs that induced or will induce a transaction in cash during this annual cycle. In this respect, accrual emission costs consist of deferring and matching outflows to the distinct period of selling the produced good. In case incurred emission costs have not yet materialised in an outlay (because the transaction in allowances to meet the obligation to surrender is still due), accruals are more informative than only observing cash flows (Kwon, 1989: 277). Unlike classical assumptions towards accrual accounting (e. g. those discussed in Christensen and Demski, 1995: 410-411), emission costs only carry one-period timing differences over a cash-basis accounting, but do not sufficiently inform on future cash flow realisations. Hence, incurring accrual emission costs fulfils the formal accounting requirements, but it still remains questionable to which extent this approach can be vindicated from an economic perspective.

If management accounting is to inform on whether an agent has chosen an action that maximises his principal's wealth or to inform on which decision yields a maximal payoff (Demski and Feltham, 1976: 63), it is uncertain whether accrual accounting procedures reveal the full picture. Considering that only a small amount of certificates was auctioned off during the first two EU ETS trading phases, it is evident that accrual cost accounting only captures those transactions in allowances that are incremental to the standard endowment. Hence, this management accounting approach fails in completely mapping the internalisation of externalities arising from emission emissions as it is originally set out by the economic theory of a trade in allowances.

This assessment of accrual cost incurrence and measurement techniques relies on the stewardship role and the decision making function of management accounting. Reliability and comparability will be considered first as characteristics attributable to both functions. Accounting numbers are perceived as reliable as long as recognition and measurement are based on observable realisations, and as long as there is little or no room for alternative treatments (Schweitzer and Küpper, 1989: 683). To which extent a firm-specific management accounting system proves to be reliable cannot be generally tied to employing accrual costs or another approach. All types of performance reporting measurement can theoretically be based on historical, current or replacement costs. In this respect, accounting preparers and users should agree in advance on consecutive valuation procedures. Historical costs and market prices are more reliable than mark-to-model or exit values (Barlev and Haddad, 2003: 406). Comparability, for its part, is understood in an inter-temporal and a material way. In this respect, both of these fundamental criteria are violated. Since the cost incurrence cannot be separated from exogenous effects like the allocation mode and a genuine "overconsumption" in emission rights (Kilger et al., 2002: 135-138), year-on-year accrual cost figures cannot be compared when the regulator chooses to tighten the legal environment. Materially the same procedures then lead to different costs.

The stewardship function is represented by the clean surplus, value creation and Pareto-efficient risk sharing conditions. Appropriate measurement techniques in accrual management accounting provide for equality in the sum of discounted cash flows and discounted earnings over a firm's lifetime. If accounting relies on forward-looking values, and hence adopts current or replacement costs, the hypothesised alignment of cash and income streams is violated (Hax, 1989: 168-169). Value creation postulates that the impact on firm value of an agent's decision is directly gauged in accounting. Generally, historical costs confine the principal's evaluation to decrements to his paid-in capital; in the case of accrual emission costs, it is the recognition criteria that distort accounts in the other direction. Since this concept waives recognising all opportunity costs induced by the scheme, the reporting system does not assign the relevant costs to the action space of an agent – e. g. to abatement measures or emission-efficient processes (Friedman and Neumann, 1980: 407-408). With incompletely described alternatives at managerial level, the possibility of an inefficient outcome for the principle looms large.

The last stewardship evaluation criterion is the efficient risk sharing postulate that holds if a risk averse agent is provided with a smooth income and hence smooth contracted compensation. Even though the ETS introduces additional components to performance reporting, accrual cost accounting reduces their influence and volatility to a minimum in the one period setting. The manager can easily predict the extent of the emission short position and hence the impact of emission costs to his compensation. Ceteris paribus, this smooth income stream does not give rise to alterations in the risk sharing arrangement (Holmström, 1979: 79). In the multi period case, though, exactly the restriction to allowance purchases adds additional uncertainty. Exogenous shocks that are uncontrollable by the manager, for instance changes in the regulatory setting or in demand, influence the required risk premium to be borne by the principal (Laux, 2006: 91). In this setting, accrual cost accounting alone is unable to resolve the agent's uncertainty in an early manner and does not control for asymmetric risk appetites (Baiman and Demski, 1980: 193).

The same issue, timely revelation of information, is the first criterion for efficient decision making, too. In this sense, not the absolute valuation impact is addressed, though, but instead reporting a relevant change in the environment when evaluating e. g. the product mix or pricing. Consequently, information economics focus on recognition than on measurement. Since the shortcomings of accrual cost accounting in value creation and risk sharing are bound to the nature of emission incurrence criteria, the same issues also extend to timely information: the fact that the underlying structure of opportunity costs has altered with the first emission trading day is only reported to the extent to which it alters current period cash flows (Demski and Sappington, 1990: 371). In this respect, accrual emission costs are timely neither for good nor for bad news.

The other characteristic of performance reporting in a decision making context is freedom from noise. Cost information collected under this approach can be generally termed as least noisy alternative, but the factual amount of noise depends on the reporting system details. With clearly set out recognition criteria, no semantic noise can be expected with respect to quantities (Lee and Bedford, 1969: 261). As to measurement, market prices should only be employed as best estimate in projecting the economic impact or future shortfalls in allowances.

This evaluation of accrual management accounting for emission allowances reveals major drawbacks of this reporting concept. Violating the value creation and risk sharing criteria (at least the latter in the multi period setting), this approach does not turn out to be suitable in case of the stewardship setting. Similarly, the radical reduction to "realised" costs also renders it meaningless in the decision making context.

3.2.3 Emission Rights as Imputed Costs[32]

3.2.3.1 Determination of Quantities

Classifying Emission Rights as Resource

As the imputed cost approach takes its starting point from an independent identification of usage, it is necessary to recall the economic theory behind certificates when adapting the reporting system. From a theoretic point of view, every allowance consumed constitutes costs – either in case of voiding purchased or granted certificates as the first gives rise to cash flows and the second to opportunity costs (Buchanan, 1999: 43; Bode, 2006: 681).

[32] This section is based on parts of Zimmermann and Veith (2007).

Economic opportunity cost considerations and management accounting theory share one common and fundamental concept that is potentially suitable for extending the recognition of emission rights in management accounting: resource allocation. Since the economic and accounting theories of resource allocation do not necessarily overlap, prior to reflecting a consumption of emission rights in performance reporting, it has to be tested whether this transaction really fulfils the characteristics of consuming a resource as defined in accounting literature. As already set out, traditional and strategic cost accounting systems theoretically are designed to inform management on resource planning (Kaplan, 1984: 405; Chenhall and Langfield-Smith, 1998: 8). Even though textbooks define costs "as a resource sacrificed or foregone to achieve a specific objective" (Bhimani et al., 2008: 38), this feature plays a subordinate role, and is usually condensed to recognising capital costs for calculating the residual income.

A management accounting extension providing for recognising emission rights can nevertheless rely on a resource consumption concept. This approach of imputing costs can be found in European management accounting systems like in Germany, Austria and Switzerland (Küpper and Mattessich, 2005: 346). Here, especially the seminal works of Kosiol (1953: 16) and Schmalenbach (1956: 6) lead to a definition of imputed costs based on resource consumption. Many authors contributed to this discussion until the 1980s, like Käfer (1964: 25), Lohmann (1964: 24), Menrad (1965: 24), Ehrt (1967: 13), Mellerowicz (1973: 6), and Heinen (Heinen, 1983: 58).

The remainder of this section sheds light on the questions if emission certificates can be defined as a resource in an accounting sense, and if these certificates are consumed in the course of production. Only having affirmatively answered these questions, an emission right is assigned costs different from zero.

An object is defined as an economic good if the following attributes hold cumulatively: suitability for purpose, command of the object, marketability, relative scarcity and equilibrium price (Kosiol, 1975: 119; Zúñiga, 1999: 302-303). To be suitable for a purpose, an object must be qualitatively appropriate in fulfilling an unmet need to increase the utility of its user (Menger, 1968: 1-3). This final relation requires the suitability of the object to be utilised in a technical and economic process to fulfil demand. Given that the production of an output – like the generation of electricity – is the economic purpose, then an emission right is an appropriate means in settling the mandated obligation of surrender, and is hence suitable to increase utility.

As long as an object like this is within the sphere of an economic agent, it is termed as available. If in addition to that, all necessary legal and physical requirements are met, the agent is provided with a feasible command of it (von Wieser, 1884: 43). Every emission right, either allocated or purchased, is at the same time characterised by availability and command due to its intangible character. However, with respect to a right it can be opposed that it does not possess a goods-character, and only comprises the access to an underlying good; in this sense, it can only be regarded as a derivative of another object (Bergner, 1967: 593-594). This objection does not convince in the case of an emission certificate that merely comprises the allowance of polluting one tonne of harmful substances. The emission itself cannot be regarded as the underlying good of a derivative since it does not increase the utility of the emitter, but is instead defined as a bad (Lehmann, 1956: 89).

An object is called marketable if it can be traded on a one-to-one basis. The latter requirement is already met by the feasible command of the object, but additionally depends on the transferability from the legal point of view (Endres, 1976: 789). By definition, an emission right is fungible (Directive 2003/87/EC, Art. 1(1)), and is traded on especially designated exchanges.

Every marketable object is defined as economic good, given its availability is scarce in relation to demand (von Wieser, 1884: 43). This ratio implies a positive value, whereas free – i. e. abundantly available – goods exhibit a value of nil. Generally, every tangible or intangible object is qualified as an economic good as long as it can be assigned a value (Nicklisch, 1932: 87). Even in case an emission right is granted, it cannot be regarded as a free good: since the allocation is perceived as a grant due to the relative scarcity of allowances, the character of an economic good is apparent (Chmielewicz, 1970: col. 1945).

Provided that market participants articulate demand for an object, buyers and sellers will agree on a price to settle the transaction. A suchlike economic good exhibits an equilibrium price in case the bid price is lower than the ask price (Kosiol, 1975: 117). Providing a bid price can be witnessed at an emission exchange, consequently demand is higher than supply. Irrespective of the type of acquisition or allocation, certificates under the current EU Emission Trading Scheme fulfil all five necessary criteria of economic goods.

Consumption and Recognition

The previous explanations showed that an emission allowance can be defined as an economic good. To include a firm's surrender of these rights in performance reporting systems, this good or resource must be consumed in the course of production. It is questionable whether allowances can be consumed at all, and economic theory does not clarify this issue. From a business level perspective, a good is consumed as soon as, by a certain transaction, it is deprived of its value-carrying character, irrespective of changes in its physical or legal nature (Kosiol, 1953: 14). Consumption comprises two variants: recognised consumption, and neutral exchange with an inflow of an equivalent counter-value. Recognised consumption happens to both, resources disappearing during the transformation, and those with an unlimited capacity (Nicklisch, 1932: 94; Heinen, 1983: 65).

Emission rights lose their suitability of an economic good once a firm surrenders certificates to the regulatory authority. For technical and physical reasons, they are not immediately consumed in the course of production like a commodity. From this point of view, the transformation process can be possibly conducted without this resource. Further, surrendering allowances cannot be qualified as a neutral exchange since any direct or indirect benefit is granted by government. As opposed to the emissions case, certain taxes fulfil the notion as lump-sum payment for drawing on public services (Koch, 1966: 52). Still, voiding certificates is a result of public coercion, and non-compliance is fined and additionally bound to surrendering the lacking amount of certificates in the subsequent trading period (Directive 2003/87/EC, Art. 16). A consumption to be recognised in management accounting hence relies on the surrender. Unlike most other rights (e. g. licences), emission rights disappear in the course of transformation. This argument does not need to be distinguished for the case of free allocation or purchase. It is only important that the obligation of transferring an economic good – here: an emission right – back to the public sector. Possibly, this transaction gives rise to consuming a good when accounting for imputed costs: in place of strictly focusing on cash flows, the relevant evaluation criterion is consumption in relation with production (Schmalenbach, 1956: 7).

Similar to cost incurrence, the realisation of revenues arising from emission rights sales draws on the conception of economic goods – however not in the sense of their consumption, but rather in the sense of a disposal on markets (Kosiol, 1958: 23). The imputed cost framework explicitly excludes realising revenues from speculative transactions since the product factor approach only focuses on those allowances being held for the intention of utilisation in transformation processes. In this respect, the incidence of opportunity costs or profits is fully revealed in management accounting. Due to the symmetric treatment of emission costs and revenues, the latter case is not further distinguished in the remainder of this analysis.

Imputation Principles

Traditionally, costs are assigned to a defined object or to a certain period during which an object is created (Kaplan, 1984: 393). This product is not necessarily restricted to a marketable economic good, but can also constitute an intermediate good that is designated to internal use. Hence, costs only incur in case of a firm is engaged in any kind of transformation process. In order to result in a management accounting approach for emission rights based on imputed costs for utilising resources, it is not sufficient to show that these rights can be qualified as an economic good disappearing in production; it has to be further explained that this consumption is related to a certain product. This idea is not associated with indicating a variable or fixed input relation, but asks for the logical relationship between costs and products. The literature on imputed cost management accounting most commonly distinguishes between a causal imputation principle and a final imputation principle (for an overview of these and other principles, see Riebel, 1994: 519-520).

The final principle posits that costs can be viewed as means employed in rendering a product (Schmalenbach, 1956: 6; Munzel, 1966: 50; Gutenberg, 1958: 132). Costs incur consuming an economic good for the purpose of production. However, this principle can be interpreted in different ways, resulting in distinct views whether consumption due to public coercion can be explained by a final relationship or not. If the aggregate production of a firm is understood as causa finalis, then every kind of consumption incurs costs as long as it is willingly accepted by the involved economic agents (Schneider, 1961: 695; Ehrt,

1967: 31). Following this broader view, surrendering emission rights always leads to imputed costs. As opposed to that, the narrow interpretation of this principle only includes willing consumption that is immediately bound to production, not providing for any form of coerced consumption (Kosiol, 1958: 26).

Equally fundamental is the causal principle; this strand of literature interprets production as cause and resulting costs as effects (Munzel, 1966: 49). Such a principle allows for a definite recognition of coerced consumption: without these costs, the respective production process would impossibly exist (Kosiol, 1958: 27). Even though some authors claim that coerced consumption cannot be recognised under a causal principle, this observation only holds for genuine cash flows like levies or taxes (Ehrt, 1967: 25). Surrendering emission rights however involves the consumption of a distinct resource that provides for recognising costs under this concept.

3.2.3.2 Measurement

Both principles, except for the narrow interpretation of the final principle, provide for an imputation of a publically coerced consumption of resources like emission rights. With which amount this transaction is finally recognised in management accounting also depends on the measurement technique employed. Since only economic goods qualify for carrying imputed costs, a positive value of the object can be assumed. Following the theory of goods, a value represents the relative subjective utility towards an object, whereas accounting objectifies this value by drawing on prices (Lehmann, 1956: 89-90). Hence the purpose of measurement is expressing this value in monetary terms to achieve comparability between different consumed resources (Kosiol, 1958: 29). Similar to accrual costs, two dominant measurement techniques in assigning imputed costs can be distinguished: historical and current costs (Schmalenbach, 1956: 191-192).

However, when relying on the consumption of resources instead of deferring cash flows, this reasoning relies on a different framework and possibly yields different results. As to stewardship, the variance analysis informs on endogenously controllable managerial actions, and hence on changes in the inventory of emission allowances. Like in the case of accrual costs, the principal will design a reporting system that fully reveals the difference between planned and (exogenously driven) prices. With this knowledge, he is able to evaluate the economic significance of variances in quantity. In the information content setting, current costs should be preferred in both, product mix and pricing decisions. In the former case, an anticipative recognition of emission allowances is implied in the regulatory goal of selecting an emission-efficient production process; in the latter case, the price effect of an ETS on the production chain must be incorporated immediately. Hence, measuring granted allowances should either be based on the market value or on a weighted average of market values during the annual cycle.

Using market values instead of historical costs is consistent with traditional maintenance of assets motivations in management accounting systems of e. g. Germany, the UK or the US (Busse von Colbe, 1989: 79; Edwards et al., 2003, 2003: 45), and can be seamlessly applied to management accounting (for a discussion see Zeff, 1962; Samuelson, 1980). Even though this school of thought advocates replacement costs and not market or fair values, current values are more suitable in the case of dynamic and volatile prices on the emission rights market: the classical argument of replacement costs supposes constantly rising factor prices due to inflation. Emission allowances as fungible commodity-like goods are characterised by price fluctuations with potentially short or medium term markdowns. The long term price for emission is bound to developments that can either serve for a persistent increase or decline. Especially technological change like the introduction of abatement technologies can induce vast long positions. Then again, political decisions to lower the cap or to costly allocate a certain percentage of rights can lead to short positions and hence a sharp rise in prices.

In this respect, accounting for emission rights at current costs serves as a good estimate for the intrinsic value of the certificate. Only in case of apparent market anomalies, the application of a weighted average method can be suggested. Employing the intrinsic value concept in management accounting measurement techniques ensures the going concern of the entity as soon as the calculated long-run lower price bound is touched.

3.2.3.3 Full Cost Accounting: Treating Emissions as Externalities

Based on imputed cost accounting, approaches like full cost accounting or environmental cost accounting emerged. Their common aim is recording costs not affecting the entity but other third party economic agents or the ecological environment. Possibly, these concepts can provide further evidence for an inclusion of emission rights in performance reporting, and are hence considered and evaluated.

A focal point of this management accounting school of thought is the definition of environmental costs. These costs embrace the production-induced consumption of resources consisting of preventing or causing environmental damage due to firm activity Frese and Kloock, 1989: 15; Lange and Fischer, 1998: 109). Considering the theory of internal and external effects, within this framework two cost types can be described: on the one hand costs already borne (i. e. internalised) by the entity, and on the other hand costs factually or temporarily shifted (i. e. externalised) to other parties (Schreiner, 1990: 201). This taxonomy is reflected in two distinct approaches. Environmental cost accounting as the first approach aims at assigning internalised environmental cost by means of separate categories and drivers (Letmathe and Doost, 2000: 426-427). To this end, cost is incurred resulting e. g. from prevention, reduction and substitution of environmentally damaging processes, from provisions, regulatory costs, taxes, levies or environmental liabilities (Roth, 1992: 113). The second approach, full cost accounting, extends the incurrence criteria (Antheaume, 2004: 444-446). Apart from traditional costs, the consumption of natural resources provided for free of charge is additionally recognised in performance reporting. In this respect, full cost accounting comprises costs incurred at firm level as well as externalised costs arising from the strain of environment (Kloock, 1993: 15).

In full cost accounting, the externalities of emissions arising from the production process have to be measured before the aggregate amount can be broken down to one unit of pollution. Existing approaches of valuing non-internalised externalities rely on (macro-) economic approaches using a mark-to-model method with barely or non-gaugeable assumptions. This induces problems from a decision-making point of view on management accounting because a measurement like this does not provide for adequately reliable data and hence introduces a noise component of unknown quantity and direction (Thurm, 1997: 331). Additionally and more importantly, there is no rationally justified decision on the choice of an appropriate valuation criterion can be met. Hence, an outcome of accounting measurement can range between the minimum cost due to public regulation and the complete internalisation of externalities, which would result in an abandonment of production (Günther, 1994: 141).

To overcome these issues, full cost accounting must be based on a suitable selection heuristic that allows for inter-subjectively and inter-temporally consistent interpretations (see for a discussion Müller-Christ and Hülsmann, 2003: 263). A reporting system aiming at the maintenance of capital is generally apt to recognise economic, non-internalised consumption that does not affect payments for an entity. With respect to an accounting system at the firm level, it is to be specified which amount of regulatory effects can be anticipated in a meaningful way (Lange and Fischer, 1998: 114). All transactions are relevant from a full cost management accounting perspective that will induce a cash outlay in the foreseeable future. By definition, this characteristic is met by all economic goods, even by those that are provided free of charge, but not by natural resources that do not constitute a marketable good.

Figure 5: Assigning Emission Costs Using Activity-Based Costing

This figure illustrates the assignment of direct costs (white colour), indirect emission costs (grey colour) and other indirect costs (black colour). The vertical bar represents the share of these costs in the drivers and object of this example.

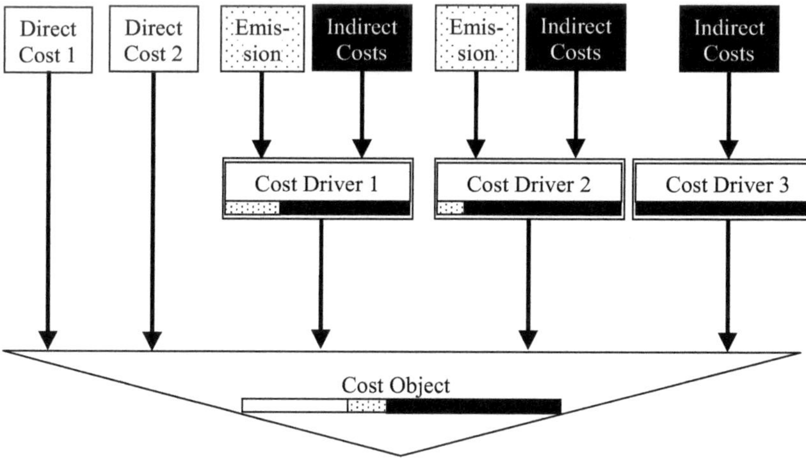

A valuation of environmental economic goods is performed at market value since a measurement at historical cost renders an internalisation of externalities impossible. This decision rule is applicable to all kinds of external effects, and is explicitly not limited to the purchase of an emission right as literature suggests (Roth, 1992: 211). In this respect, externalities – and emission allowances in particular – only incur costs if their economic relevance on firm level is likely. An approach like this is in accord with the theory of property rights and satisfies the goals of an emission trade (Dales, 1970: 95). Costs arising from compensation or emission trading schemes can hence not be regarded as penalties (Günther, 1994: 170), but as the part of an externality to be internalised.

Figure 6: Emission Rights in Imputed and Full Cost Management Accounting

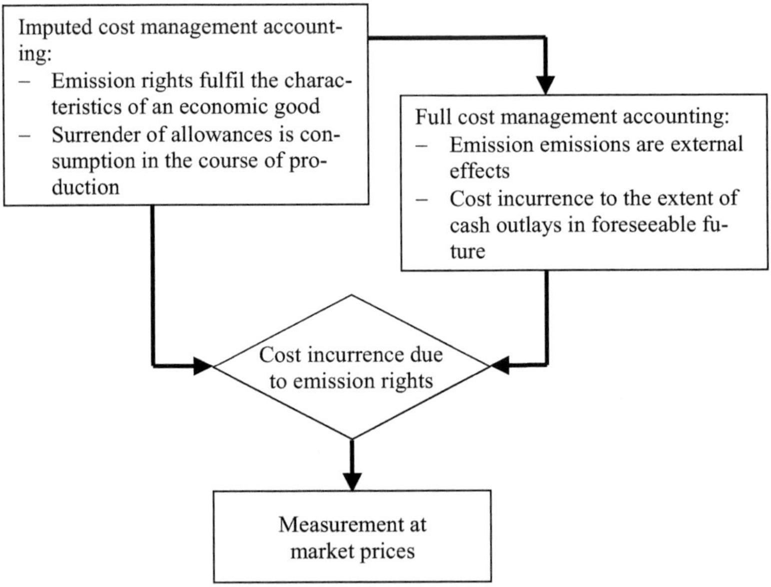

Costs for emission rights are to be traced and allocated in separate environmental accounts; a suchlike differentiation into a detached cost category discloses the trade-related burden in management reports. Figure 5 sketches the isolation of emission costs in an assignment scheme that can serve as basis to minimising the environmental and regulatory exposure in the decision process (Schaltegger and Sturm, 1992: 138). The previous discussion shows that the surrender of emission rights can be consistently mapped to management accounting if imputed costs are employed. Allowances can be regarded as economic goods without any reservation, and the type of accession is irrelevant for this classification. Further, the voiding of rights due to the legal liability corresponds to the criteria of a coerced consumption. Costs arising from this kind of consumption can be assigned under most imputation criteria. From a full cost accounting point of view, external effects arising from emission emissions are to be mapped to performance reporting to the extent of likely inducing cash outflows in the foreseeable future. An approach like this provides for cost information considering the effects of (future) regulatory actions. As depicted in Figure 6, a reliable measurement of consuming emission rights draws on market values for both freely allocated and purchased certificates.

3.2.3.4 Imputed Costs in Stewardship and Decision Making Settings

The question remains whether imputed costs yield useful information in contracting and decision making situations. Even though imputing costs for granted emission rights can be justified from a management accounting perspective in general, it can be still necessary to exclude allowances from management accounting since they do not belong to regularly acquired input goods (Koch, 1966: 61). In this respect, this section deals with specific situations in which emission rights are included in decisions in order to evaluate the presented recognition and measurement criteria as an information problem.

Assessing the general characteristics of management accounting systems for emission rights, the conclusions drawn for accrual costs apply to imputed costs with respect to reliability. Further, and as opposed to accrual costs, imputed incurrence criteria fulfil the conditions of comparability. Relying on the concept of economic goods, the goods concept meets both the inter-temporal and the material dimension of comparability. A production conforming to the regulations of the EU ETS leads to the management accounting hypothesis of consuming a resource that is independent from the employed allocation mode. Accounting users and preparers do not have to make additional provisions to provide for exogenous changes in the rule set.

Looking at the suitability of this performance reporting approach in stewardship settings starts with its compliance with the clean surplus relation. Generally, the sum of costs is not equal to the sum of a firm's cash flows if the imputed costs extend to incurrence and measurement criteria beyond cash transactions (Hax, 1989: 168-169). In other words, all freely allocated rights entering management accounting and all purchased allowances valued other than at historical costs violate this postulate. Consequently, the consumption of resources and the identity of present values concepts stand in a conflicting position. The second stewardship criterion of management accounting is value creation. Since an emission right attains its goods-character in the sense of management accounting only if it satisfies the properties of a production factor (Schmalenbach, 1956: 5; Kosiol, 1953: 16), and this concept by intention focuses on mapping transactions in the context of resource consumption and value creation. In this way, the notion of an economic good serves for defining costly input factors and as selection mechanism in reporting about future events in current reports. Management accounting as a forward-looking instrument necessarily relies on deferring and realising long-term investments by means of expected values (Dickhaut and Lere, 1983: 505). In order to conclude the assessment of the stewardship capabilities of imputed emission costs, the Pareto-efficient risk sharing is left to discuss. At first sight, the introduction of opportunity costs gives rise to volatilities in the earnings figures and hence leads to an inefficient risk sharing. With the introduction of the ETS, the manager is faced the full imputed economic effect – whereas the regulator by intention allocated large amounts of free certificates during the first two phases to allow for a soft adaption to the scheme (Åhman et al., 2007: 1718). However, apart from this one-time effect, the far-reaching character of opportunity costs reduces the uncertainties arising from regulatory change pertaining to the manager: the certificate endowment mode notwithstanding, imputed costs are systematically assigned to the respective cost object. The remaining variance in allowances consumed can be consequently eliminated by managerial action. An excessive amount of rights utilised other than due to production increases can be traced to inefficiencies or processes that emit comparatively more emission; analysing price changes separately, the unexpected component then is due to an ineffective hedge. Revealing exogenous components by means of imputed costs ideally provides for a Pareto-efficient risk sharing, as it gives a better insight to the manager's effort (Suh, 1990: 713).

The suitability of consuming emission rights in a management accounting aiming at decision making is bound to its timeliness and freedom from noise. Revealing imputed cost information as soon as possible is central to future-oriented decisions over scarce resources like process selection and pricing. In these cases, performance reports from a long-term perspective are a necessary condition for maintaining production resources, i. e. assets (Daniels, 1933: 306). Hence, incurring costs for granted allowances anticipates that, after a transition period, a distinct share of certificates will be auctioned off. Especially in product mix and pricing decisions, processes or cost objects with comparatively high emission emissions become less attractive with respect to low-emission alternatives. In this decision making context, only relying on accrual costs is insufficient. The trade-off between these two management accounting philosophies illustrates by which means the goal of altering firm level incentives to abate emission emissions becomes effective (Abadie and Chamorro, 2008: 3007-3008). Meeting the criterion of freedom from noise establishes the second component of decision-useful management accounting information. Similar to accrual costs, only a case-by-case analysis of an imputed cost accounting system can give evidence on the specific importance of noise terms. In this respect, using long-term average prices in measurement is a possible solution: it controls for noise that has an expected value of zero and additionally for mean-reverting effects of the emission price time series (Childs et al., 2001: 50-53). However, emission prices exhibited non-stationary behaviour in the past that could not be captured by relying on expected values alone (Daskalakis and Markellos, 2007: 7).

Management accounting information for long-term decisions is usually complemented by various information channels, like cash-based performance measures, or a financial accounting timely informing on downside risks in e. g. creditor situations (Watts, 2003: 210-214). The fact that information is incomplete suggests employing several instruments that are independent from each other. However, the numerical outcomes of these information channels are subject to variation since the systems with their respective information sources are only capable of approximating reality. In order to achieve a best estimate in decision making, the interpretation of several independent sources of accounting information is indispensable (Marshall, 1972: 298; Christensen and Demski, 2003: 258).

Imputed emission costs do not – and theoretically cannot – hold at once all quality criteria towards a management accounting approach for stewardship or decision making purposes. However, apart from the clean surplus relation, no characteristic can be ex ante rejected. In this respect, the resource conception in management accounting can provide for both functions. Clearly enough, reports specially designated to either function have to be generated, but instruments of both categories can be based on the same imputed data.

3.2.4 Concluding Remarks

Table 9 summarises and contrasts the economic properties of accrual and imputed emission costs. It is apparent that the selective incurrence of purchased allowances distorts the usefulness of the former approach with respect to comparability, value creation, timeliness, and – in a multi period setting – efficient risk sharing. Vice versa, the inclusion of opportunity costs due to the concept of consuming economic goods provides for a high comparability, an efficient risk sharing, and timely information.

From an imputed as well as from a full cost accounting perspective, including freely allocated and purchased emission rights in performance reporting fulfils the information needs in monitoring management and making forward-looking decisions. The analysis combining an imputed and a full cost approach illustrates domains for future research. Here, the concept of production factors is crucial that is made accessible to cost accounting via the hypothesis of resource consumptions. In determining and quantifying externalities, it provides for the selection criterion needed in full cost accounting. Ecological aspects can be meaningfully integrated into management accounting as long as consuming economic goods leads to future cash outflows with a certain probability. This provision implies that not any arbitrary amount of granted certificates can be defined as costs: the further away a future period induces an uncertain economic burden, the more conservative should be the inclusion practice of these economic goods into accounting. An anticipation of financial outflows that can hardly be justified leads to decoupling management accounting from any accruals- or cash-based information system. Controlling for this effect, the imputed cost accounting still is not deprived by its abstraction from cash flows, as opposed to accrual cost accounting.

Table 9: Assessment of Management Accounting Approaches for Emission Rights

Function	Characteristic	Accrual costs	Imputed costs
Common Characteristics	Reliability	*Yes/No* Possible, but depends on management accounting specific arrangements	
	Comparability	*No* Cost incurrence depends on exogenous effects	*Yes* Universal comparability due to concept of economic goods
Stewardship	Clean surplus	*Yes/No* Dependent from valuation: historical costs hold clean surplus condition, current or replacement costs do not	*No* Inclusion of granted rights violates clean surplus condition
	Value creation	*No* Incomplete monitoring of changes in values: disregard of opportunity costs	*Yes* Consumption of resources incorporates all changes in value
	Pareto-efficient risk sharing	*No* Cost incurrence dependent on exogenous effects in the multi period case	*Yes* Low uncertainty about future external shocks by incurring costs for granted and purchased rights
Decision making	Timeliness	*No* Selective incurrence criteria do not fully reveal information	*Yes* Consumption of resources incorporates information in a timely way
	Freedom from noise	*Yes/No* Possible, but depends on management accounting specific arrangements	

3.3 Consequences for Financial Accounting

3.3.1 Functions of Financial Accounting

Similar to performance reporting, financial accounting rules also need to be interpreted or adapted to allow for the mapping of emission rights. Emerging trading schemes in the US and EU have created a demand for solutions within the financial accounting framework since regular reports are legally compulsory at least for listed firms (Zimmermann et al., 2008: 11). Due to the complex economic nature of an ETS, roughly half a dozen of specific treatments have been proposed in literature or adopted by regulatory bodies. Following the previous analysis of management accounting, the available alternatives will be discussed pursuing a positive approach. The underlying intuition is simple – if financial accounting is meaningful, it will cause economic consequences for related parties (Zeff, 1978: 56). In this vein, two major functions of financial accounting can be found: on the one hand the supply of information exhibiting decision usefulness and on the other hand the supply of information being suitable in contracting (Holthausen and Watts, 2001: 33, 49; Wagenhofer and Ewert, 2007: 4). It is apparent that these two functions are (nearly) identical with those outlined for management accounting. Differing is the set of intended addressees, since financial reports are intended to be used by i. a. investors, creditors, customers, governments or the public (IASB FW.9). Consequently, the information need articulated by the respective stakeholder group determines what is more precisely meant by e. g. "decision-useful" information.

It is conventional wisdom that US scholars emphasise the information function of financial accounting for the sake of equity valuation purposes (Holthausen and Watts, 2001: 24). A rational economic agent bases a decision to invest into a firm on the expected value of future cash flows, and consequently needs to draw on information like financial accounting to estimate them (Kothari, 2001: 108-109). Investors on a capital market act as outsiders as they do not dispose of the full information set available to the managers. To mitigate this agency conflict, investors ask for (audited) accounts that can serve as an unbiased proxy for the economic profit of the firm (Christensen and Demski, 2003: 334-335). This implies two theoretical characteristics of information efficient financial accounting: Firstly, financial accounting is to report on all economic events that can be reliably measured and sufficiently verified (Ball et al., 2000:

6). Secondly, managers must not be able to opportunistically exert discretion over the report (Barth et al., 2008: 471). Financial accounting rules in force necessarily deviate from these assumptions for timing, transaction cost and agency reasons.

The contracting function of financial accounting does not necessarily draw on individually arranged contracts but can also root in legal provisions in the interest of a related party. For example protecting creditor rights can be accomplished by means of negotiated lending agreements (so called debt covenants used in the US, see e. g. Begley and Feltham, 1999) or by legal rules restricting dividend payouts (to be found e. g. in Germany, see Schneider, 1994: 38). Creditors and investors have conflicting interests regarding the redistribution of capital earned in entities with limited liability. Shareholders are residual claimants and bear economic incentives to maximise the dividend yield. Bondholders face an asymmetric payout and vote for minimising cash outflows to maintain their collateral for the case of default (Jensen and Meckling, 1976: 337-338; Watts, 2003: 211). In a related way, the contracting function of financial accounting also extends to stewardship. Bonus plans do not necessarily rely on performance reports but can be tied to financial accounting since its income is more readily available to outside investors. Both functions, debt and compensation contracting, give rise to similar characteristics of accounting numbers: to understate net assets and earnings (Watts, 2003: 213). This deviation between economic and accounting realities can be either accomplished in an arbitrary manner, being termed unconditional conservatism (Ball and Shivakumar, 2005: 89-90), or requiring a higher degree of verifiability for accounting earnings than losses, giving rise to conditional (Basu, 1997: 6-7). Conditional conservatism conveys more efficient information in lending and bonus contracts since it incorporates an adequate signal to accounting income informing those parties interested in the downside risk (Ball et al., 2008: 177).

The dichotomy in functions also extends to the valuation rules in financial accounting. In determining accounting rules, two concepts compete for primacy: the revenue/expense and the asset/liability approach (Samuelson, 1996: 148).[33]

---

[33] In the German literature, terms like dynamic balance sheet and static balance sheet are used for the revenue/expense and asset/liability view. Apart from this difference, both approaches conceptually imply the same.

The former one postulates to recognise revenues in the period when the economic inflow is sufficiently certain, and to match related expenses accordingly. It promotes the income statement as information instrument since the balance sheet is considered to capture all temporary items not yet released to profit or loss (Schmalenbach, 1962: 72). The latter one relies on the balance sheet as primary device in reporting and defines assets (liabilities) as giving rise to future cash inflows (outflows). Already Simon (1899: 304) defined the value of an asset as its "distinct utilisation or exit value". Consistently, the income only maps changes in stocks. Not accidentally there exist two financial accounting and valuation functions: As the revenue/expense view focuses on the verifiable accomplishments of the management, it is closely linked to the contracting function (Ahmed, 1983: 300). Attempting to measure the inherent value of assets based on updated information, the modern asset/liability approach relies on fair values (Ronen, 2008: 184). Since this measure is often proxied by market values, it is in line with the information function of financial accounting. Recent discussion papers and exposure drafts issued by the IASB as well as by the US American Financial Accounting Standards Board (FASB) are no longer coined by the revenue/expense but by the asset/liability view (Benston et al., 2006: 172).

In the remainder of Section 3.3, the most important financial accounting alternatives for emission rights will be described and evaluated with respect to their conformity to either school of thought (i. e. functions or valuation conceptions). Apart from this normative approach, in Section 4.7 the usefulness of the information processed by selected accounting treatments is evaluated in a market valuation setting by means of empirical and modelled data. This offers a direct test for the suitability with the decision usefulness paradigm.

3.3.2 The Accounting for Emission Rights Debate

With the start of the US Acid Rain Program as a part of the Clean Air Act Amendments of 1990, the Federal Energy Regulatory Commission pronounced the first financial accounting rules for emission rights. They are codified in the

Uniform System of Accounts (USofA)[34] 101.21 and are applicable under the framework of US Generally Accepted Accounting Principles (GAAP). The earliest contributions to literature regarding the question of how to map emission rights in accounting are rooted in the critical perspectives literature and rebuke this rule set. Gibson (1996: 658) and Lehman (1996: 74) reject any form of recognising allowances in accounting since using the environment as a sink can neither be proxied by financial items nor can accounting for it give rise to economic incentives to lower pollution. The first technical papers dating from the same year acknowledge a function for accounting in reporting over the firm-specific exposure towards a trading scheme but propose amendments to USofA (Milne, 1996; Wambsganss and Sanford, 1996).

The introduction of the European trading scheme created a demand for an accounting treatment in line with IFRS as the compulsory norms for all EU-based listed firms. The IASB decided not to issue a separate standard but to rely on an interpretation of effective rules for this special accounting problem. Consequently, the International Financial Reporting Interpretations Committee (IFRIC) as designated body was assigned to develop an interpretation. In 2003-05, a first draft had been presented, being mainly an interpretation of the International Accounting Standard (IAS) 20 (Accounting for Government Grants and Disclosure of Government Assistance)[35], IAS 37 (Provisions, Contingent Liabilities and Contingent Assets) and IAS 38 (Intangible Assets). It attracted a considerable number of objecting comment letters during the subsequent due process: out of 39 letters received by national rule setters, auditing firms or industry associations, 31 rejected the proposal (Völker-Lehmkuhl, 2006: 79-84). Main reason for the controversy is the inconsistent valuation techniques applied for the arising balance sheet items capturing the allowances, the allocated grant and the emission liability creating an accounting mismatch. Nevertheless, in 2004-12 the final version of IFRIC 3 (Emission Rights) was adopted by the IASB (Zimmermann, 2006: 369).

---

34 Uniform System of Accounts Prescribed for Public Utilities and Licensees Subject to the Provision of the Federal Power Act, codified at 16 USC 791a et seq.

35 All IFRS named in this study refer to the legal status as endorsed by the European Union in 2008, published in Regulation (EC) No. 1126/2008 of 3 November 2008 adopting certain international accounting standards in accordance with Regulation (EC) No. 1606/2002 of the European Parliament and of the Council, OJ L 320: 1-565.

Before an IFRS or IFRIC can attain legal force in the EU, it is subject to a comitology process conducted at several European bodies. At a first stage, the Commission requests the European Financial Reporting Advisory Group (EFRAG) to assess the technical aspects of any new standard or interpretation adopted by the IASB. EFRAG is a private sector institution set up by the Commission and funded by the accounting profession and industry associations (Luthardt and Zimmermann, 2009: 83). At a second stage, the standard or interpretation is endorsed by the Accounting Regulation Committee (ARC). It consists of delegates representing each Member State and asks for the political legitimacy of the regulation (Chiapello and Medjad, 2009: 454). After both bodies have recommended the document, it enters the political process and has to be approved by the European Council and Parliament. IFRIC 3 did not receive the approval of either committee in 2005-05: EFRAG recommended its rejection, and the ARC proposed to delay its first mandatory application (Völker-Lehmkuhl, 2006: 73-74). Not only being faced with the criticism of practitioners but also risking a failure also in the political process of the endorsement, the IASB decided to withdraw IFRIC 3 in its 2005-06 session (Lorson and Töbe, 2008: 500). Despite its withdrawal, it can be (and empirically is) still applied as it only interprets underlying standards in force. The withdrawal of IFRIC 3 did not make the proposal mandatory. That is why, a large number of recent technical articles scrutinises IFRIC 3 and describes its inconsistent valuation techniques (e. g. Hermes and Jödicke, 2004; Günther, 2006; Völker-Lehmkuhl, 2006; Zimmermann, 2006; Lorson and Töbe, 2008; MacKenzie, 2009; Rogler et al., 2009).

The accounting for emission rights debate has not ceased with the withdrawal of IFRIC 3. Currently, the IASB and FASB prepare a joint accounting standard for Emission Trading Schemes. The first exposure draft is scheduled to be released in the second quarter of 2010, and the final standard is envisaged to be in force by the first half of 2011. It is planned to present an integrated solution for compulsory and voluntary schemes that can be operated either as a cap-and-trade or as a credit-based system. Consequently, more recent contributions discuss which issues need to be resolved for a refined accounting solution (Cook, 2009). Some authors note that financial disclosures may not be sufficient to fully understand the ETS-related implications (Bebbington and Larrinaga-González, 2008; Kolk et al., 2008), and point at the relevance of more directly

ETS-related but non-financial information like allocation and compliance reports.

Finally, three empirical studies have so far estimated the value relevance of ETS-measures in an accounting setting. These articles only focus on the US Acid Rain Program and do not assess the capital market properties of accounting treatments for emission rights in a direct manner. Hughes (2000: 221) finds a negative impact of a non-financial pollution proxy during the adoption of this trading scheme. Similarly, Hughes and Reynolds (2001: 1368) estimate a higher earnings response coefficient for polluting firms during the same time period. In the contribution of Johnston et al. (2008: 757), the number of allowances held turns out to be overall positively valued by investors.

3.3.3 Accounting Alternatives

3.3.3.1 Major Treatments

Case Study Set Up

Due to the complex nature of an ETS giving contemporaneously rise to economic assets and liabilities, the major alternatives having emerged to map these issues in financial reports will be illustrated by means of a case study. Since the future analysis focuses on the consequences of the EU ETS on capital markets, only alternatives with relevance for EU-based listed firms are discussed. This extends firstly to the withdrawn IFRIC 3 that outlines two different alternatives for subsequent valuation: a cost and a revaluation model. Additionally, an amendment to IAS 38 as proposed by the IFRIC is presented envisaging to measure allowances at full fair value (IASB, 2005). Because no specific international standard exists, IFRS firms may invoke IAS 8.12[36] and use accounting policies from other standard setters. That is why the methods applicable under US GAAP, USofA 101.21, are described here, and as they are often employed by EU firms they can be considered a final major treatment. The approaches differ broadly in terms of recognition, measurement, and effects on income over time.

For the sake of comparability, this case study follows the illustrative example of IFRIC 3 as presented by Cook (2009: 10-12), but additionally introduces

[36] IAS 8 (Accounting Policies, Changes in Accounting Estimates and Errors).

the effects on quarterly reports. Since the EU ETS clears the obligation and assets for a calendar year on the subsequent April 30, this case study includes all accounts of until the second quarter of the adjacent fiscal year to portray the full effect of one trading cycle. Initially, the company is granted 12,000 emission rights at a price of 10 EUR/t. The price for allowances quotes at 11 EUR/t at the end of the first quarter, at 12 EUR/t in the second and third quarter and drops to 11 EUR/t at year end. The emissions accrue to 2,750 t in each of the first two quarters and to 3,500 t in each of the last two quarters. For the first half of the subsequent fiscal year, the new emissions pertaining to the following year and their certificate price changes are disregarded to isolate the ETS effect of the yet to be completed cycle. The shortfall in 500 certificates is levelled on December 31.

IFRIC 3 Cost Method

The basic idea of IFRIC 3 is the recognition of all assets and liabilities related to an ETS. It therefore specifies three separate balance sheet items: an intangible asset, a deferred income and a liability. Both the cost and revaluation methods of IFRIC 3 follow the same rules of initial measurement applicable to these three accounts. In line with them, the asset reflects all purchased and granted certificates that are both initially measured at fair value (IFRIC 3.6). The fact that even certificates allocated for free are to be fully recognised is by justified by the IFRIC that they equally fulfil the requirements set out in IAS FW.53-58 and constitute an asset in general. More specifically, the free allotment fulfils the criteria of IAS 38.8 and 21 applying to intangible assets: both the potential to generate future economic benefits and the trade on an active market are independent of the type of acquisition (see also Hermes and Jödicke, 2004: 291). Additionally, allowances received at no charge are interpreted as a non-monetary government grant (IAS 20.23). A deferred income component is recognised at the initial fair value of a received grant and provides for an income-neutralising treatment of allocated rights, at least at the outset of the trading period. IFRIC 3.7 explicitly excludes the net approach of IAS 20.27 to recognise these allowances at the amount paid for them as this treatment does not fairly represent the assets received (Zimmermann, 2006: 370). Finally, IFRIC 3.7 prescribes to defer a provision capturing the monetary value of occurred physical emissions. Following IAS 37.14(a) and 17(a), the liability only extends to past events creating a legal obligation; hence at allocation date, no provision is to be made to reflect e. g. the expected amount of emissions.

According to the cost model, the allowance asset is measured at historical cost after recognition (IAS 38.72 and 74). Assuming a constant number of rights held but a volatile price for emissions, an asymmetric valuation is introduced due to the impairment only approach (IAS 36.59[37]): it leads to reporting expenses for decreasing certificate prices following the lower of cost or market rule, but no corresponding revaluations are found in the opposite case with increasing quotations. A hidden reserve on the asset is only revealed when it is disposed of (Rogler, 2005: 260). The government grant cannot be revaluated; instead, it is systematically amortised according to the portion of usage, i. e. incurred emissions. Changes to the liability arise if the best estimate regarding the amount needed to settle the obligation is revised (IAS 37.36). Transposed to the emission case, IFRIC 3.6 defines to calculate its value by multiplying the necessary number of allowances with their market price (Rogler et al., 2009: 374). Increases over time occur as emissions and/or prices rise, but in case of dropping prices, the increase in the liability is ceteris paribus more slowly, and can be even negative in extreme cases. It is apparent that this diversity in measurement rules hampers a straightforward interpretation and can give rise to inconsistent effects to earnings.

Table 10 illustrates the consequences of the IFRIC 3 cost method to balance sheet and income statement when applying the outlined example. The only factual cash outlay amounts to 5,500 EUR and occurs in the fourth quarter. During the first interim period, 2,750 t are emitted that – measured at the closing price of 11 EUR/t – give rise to an obligation of 30,250 EUR, which is also recognised in income. The effect of risen allowance prices is neither reflected in the asset (due to the lower of cost or market rule) nor in the deferred income. Hence, the expenses on the liability measured at current cost cannot be completely compensated by derecognizing the matching share in the deferred income, resulting in a net loss of 2,750 in the first quarter. Analogous considerations apply to the two consecutive quarters. In the fourth quarter, the increase in the emission liability is smaller than derecognising the grant. This arises from the differ-

[37] IAS 36 (Impairment of Assets).

ent measurement techniques used for the liability (current cost) and the grant (systematic amortisation), as well as from the drop in emission rights prices (from 12 to 11 EUR/t). At the date of derecognizing all related balance sheet items, i. e. on April 30 in the second quarter of the consecutive year, the hidden reserve on the allowance asset (10 EUR/t at recognition, 11 EUR/t at the end of the trading cycle) is revealed and realised to the profit or loss account. The net retained earnings of 5,500 EUR equal the cash outlay needed to settle the excess pollution in the fourth quarter of the fiscal year. Such a timing difference as well as the volatile interim earnings mark the major issue of this treatment: the measurement or reporting mismatch (Günther, 2006: 234) resulting from an inconsistency between the asset (here at cost), the deferred income (amortised at cost), and the liability (at current value).

Table 10: Sample Accounts Conforming to IFRIC 3 Cost Method
(adapted from Cook, 2009: 466)

	1st Qrt.	2nd Qrt.	3rd Qrt.	4th Qrt.	1st Qrt.	2nd Qrt.
Balance Sheet						
Assets						
Allowances	120,000	120,000	120,000	125,500	125,500	0
Cash	0	0	0	-5,500	-5,500	-5,500
	120,000	120,000	120,000	120,000	120,000	-5,500
Liabilities						
Obligation	30,250	66,000	108,000	137,500	137,500	0
Grant	92,500	65,000	30,000	0	0	0
Equity						
Retained Earnings	-2,750	-11,000	-18,000	-17,500	-17,500	-5,500
	120,000	120,000	120,000	120,000	120,000	-5,500
Income Statement						
Gain on Disposal	0	0	0	0	0	12,000
Amort. of Grant	27,500	27,500	35,000	30,000	0	0
Ch. in Obligation	-30,250	-35,750	-42,000	-29,500	0	0
Profit (+) / Loss (-)	-2,750	-8,250	-7,000	500	0	12,000
Non-Financial Emission Information						
Allowance Price	11.00	12.00	12.00	11.00	11.00	11.00
Allowances Held	12,000	12,000	12,000	12,500	12,500	0
Cumul. Emissions	2,750	5,500	9,000	12,500	12,500	0

IFRIC 3 Revaluation Method

Compared to the cost model of IFRIC 3, the revaluation method leaves all previously explained parameters unchanged except for the subsequent measurement of the allowance asset. This option stems from IAS 38.72 and 77 allowing a revaluation of intangible assets. Impairments give rise to costs whereas asset appreciations accrue in the revaluation reserve (Zimmermann, 2006: 371). In the latter case, the model operates under a dirty surplus regime since the profit and loss account remains untouched for revaluations. Two consequences emerge: firstly, at any time the balance sheet reports the fair value of the emission right asset which can be considered as finer information. Secondly, the earnings figure largely remains unchanged; the only difference to the cost model can be found in asset derecognition when no hidden reserves are revealed.

Table 11: Sample Accounts Conforming to IFRIC 3 Revaluation Method
(adapted from Cook, 2009: 467)

	1st Qrt.	2nd Qrt.	3rd Qrt.	4th Qrt.	1st Qrt.	2nd Qrt.
Balance Sheet						
Assets						
Allowances	132,000	144,000	144,000	137,500	137,500	0
Cash	0	0	0	-5,500	-5,500	-5,500
	132,000	144,000	144,000	132,000	132,000	-5,500
Liabilities						
Obligation	30,250	66,000	108,000	137,500	137,500	0
Grant	92,500	65,000	30,000	0	0	0
Equity						
Reval. Reserve	12,000	24,000	24,000	12,000	12,000	12,000
Retained Earnings	-2,750	-11,000	-18,000	-17,500	-17,500	-17,500
	132,000	144,000	144,000	132,000	132,000	-5,500
Income Statement						
Amort. of Grant	27,500	27,500	35,000	30,000	0	0
Ch. in Obligation	-30,250	-35,750	-42,000	-29,500	0	0
Profit (+) / Loss (-)	-2,750	-8,250	-7,000	500	0	0
Non-Financial Emission Information						
Allowance Price	11.00	12.00	12.00	11.00	11.00	11.00
Allowances Held	12,000	12,000	12,000	12,500	12,500	0
Cumul. Emissions	2,750	5,500	9,000	12,500	12,500	0

As Table 11 shows, the revaluation method leads to accruing more total assets due to the overall allowance price increase. It can be argued that leaving all other things unchanged, inflating total assets deteriorates performance measures like return on assets, but at the same time the larger regulatory asset makes the (quantitatively equivalent) ETS-related profits or losses relatively less voluminous. The effect of accruing the revaluation reserve can be particularly observed throughout the first two quarters of this example. Because of the neutral asset derecognition, the net retained earnings are not affected after fiscal year-end. Consequently, compared to the example earnings impact of the cost method, only the very last interim period is different as there is no disposal effect.

Proposed Full Fair Value Method

As early as in 2003-12, IFRIC proposed amending IAS 38 and introducing a third category of intangible assets (IASB, 2008: para. 11). The suggested amendment would allow recognition of all changes in the asset's fair value directly through profit and loss, reducing mismatches among the three balance sheet items (emission rights asset, carbon liability, government grant) and the impact of their changes on earnings. Apart from that, the approach of IFRIC 3 remains unchanged, and there would be no consequences for measuring and valuating the emission liability and the government grant. Even though this treatment matches the measurement of the carbon asset and liability, the underlying quantities for the allowances held and the emissions discharged still differ. Moreover, the measurement of the government grant remains untouched so that the measurement and reporting mismatches with respect to the liability prevail. The proposal does not advance any new approach for the deferred income as IAS 20 is currently under revision.

Since this proposed treatment only alters the realisation rules, the amount of total assets is the same as in the IFRIC 3 revaluation method case. Table 12 sheds light on the accounting effects of this full fair value proposal. This approach results in a net financial year income impact equalling the outlay for the additional allowances, here 5,500 EUR. Still, the ETS-related interim income remains volatile during the emission cycle and it can be questioned whether the disclosure fairly represents the economic exposure towards the scheme. The fact that the recognition and measurement rules of the asset, liability and deferred income are not aligned, a rising price for emission rights leads to emission profits and vice versa.

Table 12: Sample Accounts Conforming the Proposed Full Fair Value Method
(adapted from IASB, 2005: 7)

	1st Qrt.	2nd Qrt.	3rd Qrt.	4th Qrt.	1st Qrt.	2nd Qrt.
Balance Sheet						
Assets						
Allowances	132,000	144,000	144,000	137,500	137,500	0
Cash	0	0	0	-5,500	-5,500	-5,500
	132,000	144,000	144,000	132,000	132,000	-5,500
Liabilities						
Obligation	30,250	66,000	108,000	137,500	137,500	0
Grant	92,500	65,000	30,000	0	0	0
Equity						
Retained Earnings	9,250	13,000	6,000	-5,500	-5,500	-5,500
	132,000	144,000	144,000	132,000	132,000	-5,500
Income Statement						
Remeas. of Allces.	12,000	12,000	0	-12,000	0	0
Amort. of Grant	27,500	27,500	35,000	30,000	0	0
Ch. in Obligation	-30,250	-35,750	-42,000	-29,500	0	0
Profit (+) / Loss (-)	9,250	3,750	-7,000	-11,500	0	0
Non-Financial Emission Information						
Allowance Price	11.00	12.00	12.00	11.00	11.00	11.00
Allowances Held	12,000	12,000	12,000	12,500	12,500	0
Cumul. Emissions	2,750	5,500	9,000	12,500	12,500	0

USofA 101

USofA 101.21 is the applicable reporting treatment for emission rights for US-based utilities. Even though the US and the EU trading schemes address different emitted substances and the political and regulatory backgrounds differ, the accounting issue is qualitatively the same. The idea behind the rule set is fairly simple: allowances are accounted for at cost, an asset is only recorded for purchased emission certificates, and certificates which are allocated for free are not reflected in the financial accounting. This means that certificates are only recognised when they are held beyond the endowment either for trading purposes (classified as other investment) or for settling an obligation that is higher than the current regulatory goal (classified as inventory). The price of the market transaction is the initial basis for valuation. During the trading year, purchased allowances are accounted for at cost and amortised in accordance with their usage (Völker-Lehmkuhl and Lösler, 2005: 460). If, however, an entity does not hold the estimated required amount of emission certificates, a liability reflecting the number and current price of missing rights must be recognised (USofA 101.21(F)). Thus, there is an impact on earnings only if a firm has to purchase additional allowances, deferring the expense to the time when the emission rights are used. The risk of an accounting mismatch between assets measured at cost and the liability at current value is virtually inexistent.

Table 13: Sample Accounts Conforming to USofA 101

	1st Qrt.	2nd Qrt.	3rd Qrt.	4th Qrt.	1st Qrt.	2nd Qrt.
Balance Sheet						
Assets						
Allowances	0	0	0	0	0	0
Cash	0	0	0	-5,500	-5,500	-5,500
	0	0	0	-5,500	-5,500	-5,500
Liabilities						
Obligation	0	0	0	0	0	0
Equity						
Retained Earnings	0	0	0	-5,500	-5,500	-5,500
	0	0	0	-5,500	-5,500	-5,500
Income Statement						
Amort. of Allces.	0	0	0	-5,500	0	0
Ch. in Obligation	0	0	0	0	0	0
Profit (+) / Loss (-)	0	0	0	-5,500	0	0
Non-Financial Emission Information						
Allowance Price	11.00	12.00	12.00	11.00	11.00	11.00
Allowances Held	12,000	12,000	12,000	12,500	12,500	0
Cumul. Emissions	2,750	5,500	9,000	12,500	12,500	0

Table 13 illustrates the materially different net approach as applied according to USofA 101.21. Allowances which are allocated for free are not reflected in the financial accounting, and accordingly, the first three quarters of the fiscal year do not inform on the firm's participation in the ETS. In the fourth quarter, the necessary allowances are purchased for 5,500 EUR. Since there is no shortfall to be reported, no liability is recognised, and the allowances are directly amortised, i. e. at the assumed time of usage.

Comparison of the Major Treatments

The underlying rules of the four accounting alternatives are summarised in Table 14. It is apparent that the IFRS-based treatments do not differentiate between the types of acquisition in rights, whereas the US GAAP approach only informs on purchased units. This is also reflected in the nature of the liability that in the former case represents the market value of all accumulated emissions and in the latter case captures the amount of allowances presently being short. Apart from these issues related to the balance sheet presentation, the metrics applied in the income statement are equally disjointed: USofA assumes that allowances are used, with purchased ones being reflected in the profit or loss account. Suffering from mismatches between all three balance sheet items, the net effect on earnings hardly follows any accounting intuition in the case of IFRIC 3.

USofA and the IFRIC 3 alternatives can be easily attached to valuation concepts. USofA follows the revenue/expense view as the balance sheet items are only accrued to serve for timing the recognition of accounting earnings. As profits are designed to usually arise on the sale of excessive rights and losses capture the (future requirement to) purchase additional units in case of a shortage, the income statement is the central information instrument. By means of the simple concept of netting, a relevant signal for contracting purposes is created that reports how well the managers comply with the regulation. At further inspection, this disclosure may be too simplistic in the stewardship setting since a mere "long"/"short" information does not map the complete underlying economic effect of an ETS to business. It is e. g. unable to distinguish the business-as-usual case from the case in which investments or production reductions are required to attain the emission target, with the latter states induce lower net cash inflows than the disclosed regulatory costs. From a creditor's perspective, USofA gives rise to vast hidden reserves since allocated allowances remain undisclosed. Even

Table 14: Comparison of Alternative Accounting Treatments

	IFRIC 3 Cost	IFRIC 3 Revaluation	Full Fair Value	USofA
Panel A: Purchased Allowances				
Classification	Intangible asset			Inventory
Initial measurement	Fair value at time of purchase			
Subsequent measurement	Cost less impairments	Fair value		Cost less amortization of "used" allces. and impairments
Panel B: Allocated Allowances (Free of Charge)				
Classification	Intangible asset as well as deferred income for government grant			Inventory
Initial measurement	Fair value at time of allocation			Cost (= 0 for grants)
Subsequent measurement	Cost less impairments	Fair value		
Panel C: Obligation				
Classification	Liability			
Initial and subsequent measurement	Fair Value (market price of current accumulated emissions)			Fair Value (market price of shortfall in allowances)
Panel D: Income Statement Effects				
Subsequent measurement of allowances	None if prices increase; depreciation if prices decrease	None if prices incr. (recogn. in equity), depreciation if prices decrease (if exceeding rev. reserve)	Gains if prices increase, losses if prices decrease	None if prices increase; depreciation if prices decrease
Physical emissions	Increment in liability			N/A
"Usage" of allowances	N/A			Amortization of "used" purchased allowances
Subsequent measurement of liability	Expense (rising prices) or gains (falling prices)			
Grant	Income is recognised on a systematic basis over the useful life, changes in allowances' market value are not recognised			N/A

though emission rights regularly lose their value at the end of a respective trading cycle, certificates represent a highly liquid asset class irrespective of their type of acquisition. This is consistent with unconditionally understating total assets, but at the same time the accountant's impairment possibilities on the arrival of bad news is restricted. From a normative point of view, USofA is designed to suit the contracting function better than the information function of financial accounting. Due to the coarse, netted signal, it is not ideal to serve the parties' information needs. IFRIC 3, for its part, can be attributed to the asset/liability school of thought, as the allowance asset and the emission liability fully reveal the underlying economic effects. A relic from the revenue/expense approach is the government grant that is designed as deferred income; it does not fit into the gross approach of this treatment. If the hypothesis is accepted that investors are more strongly interested in the representation of accounting stocks than in the one of flows, then the revaluation model of IFRIC 3 discloses near fully the demanded information. Taking into consideration the comments of accounting scholars and practitioners on the prevalent mismatches, it can be concluded that the income statement has not (yet) lost its relevance, even for investors. This issue is still more meaningful for the contracting function of accounting as it focuses on the properties of reported income: neither compensation nor debt contracting is put forward as long as the noisy signal suffering from mismatched revaluation streams is in place.

3.3.3.2 Further Treatments and Recent Proposals

Since the withdrawal of IFRIC 3, a handful of other proposals were made in the literature and during discussions on new accounting treatments for emission allowances. Three of them will be briefly presented in this section, before outlining the recent proposals with respect to the future accounting standard Emission Trading Schemes. To begin with, the government grant approach should be named as proposed amendment to IFRIC 3. It embraces an allowance asset and a corresponding deferred income (capturing a potential government grant), both initially measured at fair value and subsequently measured at cost. However, no amortisation of the grant takes place until it is eventually derecognised when surrendering the certificates at the end of a trading period. Consistent with US

GAAP, a liability is only deferred for the amount of units additionally required (Lorson and Többe, 2008: 503).

One of the most prominent proposals is the so called unit-of-pollution method that was advanced in a comment letter by EFRAG. This method again assumes an intangible emission right asset and a government grant due to the initial measurement of allocated certificates at fair value. In the subsequent periods, both items are amortised according to the units used, or are impaired as the lower of cost or market rule holds (Hommel and Wolf, 2005: 320). As long as the firm is sufficiently endowed, no net earnings impact emerges since both amortisation streams are matched. A shortfall is reported by means of a net liability whose changes provide for net expenses (EFRAG, 2003).

As a third alternative, a hedge accounting method was suggested by EFRAG (IASB, 2005: paras. 29-34). Accordingly, it is assumed that allowances are the only currency to settle the arising emission obligation. In this respect, the designated rights are used to hedge the cash flows related to forecast emissions. The balance sheet items as employed by the IFRIC 3 revaluation model all remain in place and the rules for subsequent measurement are the same for the liability and grant. Only the remeasurement of the asset differs since the difference between the allowances at hand and required defines whether the hedge is effective or not. If sufficient certificates are held to cover the increment in emissions, no profit or loss is realised: a temporary difference arising from the accounting mismatch between the three items is recycled, i. e. accrued to equity (Glaum and Klöckner, 2009: 330). At the end date of an effective hedge – here: when the emission rights are to be surrendered –, the net effect of equity changes is reversed. In case of a shortfall, the hedge is not fully effective, resulting in an immediate recognition as loss. In sum, the effect on the income statement is the same as under e. g. USofA. The underlying rules are more complex than this simplifying description suggests; hence, this proposal has been criticised since not every regulated entity is able to operate hedge accounting.

Formulating the future accounting standard Emission Trading Schemes attempts to resolve all technical issues voiced over IFRIC 3. The IASB still favours an initial measurement of allowances at fair value acknowledging their nature as an asset (IASB, 2009: para. 23). This raises the question of how to account for the emerging government grant; since IAS 20 is currently under review, a consistent solution in line with the future provisions of accounting for grants in general is aimed at. In doing so, three approaches are discussed that provide for a neutral recognition of allocated rights out of which the performance obligation model is recommended by the IASB staff (IASB, 2009: para. 113). The underlying intuiting is twofold: An obligation element assumes that receiving granted certificates is equivalent to receiving a loan. Further, a performance component draws on the physical amount of emissions made – if they exceed the loan conferred, additional rights need to be purchased and vice versa. This gives rise to a liability amounting to the initially recognised granted allowances that is released as profit at the end of the trading period (IASB, 2009: paras. 67-70). These provisions are said to be similar to IAS 41.35 prescribing to measure a government grant related to agriculture at its fair value. Currently, the deliberations on the forthcoming accounting standard have not yet touched any other of the further open questions. This extends to the subsequent measurement of the allowance asset that is discussed to be either measured at cost, at full fair value or split to both models either for allowances held to settle the obligation or for trading (IASB, 2009: para. 116).

Additionally, it is unresolved of how to design the emission liability. It is discussed to define an increasing liability capturing all accumulated emissions, compensated by a corresponding reduction in the grant. Conversely, when the liability is set out only to inform on the endowment status it can be deferred as soon as the shortfall becomes apparent or gradually over the trading cycle. Measurement is envisaged to be at current cost or tied to the applied measurement rule of the allowance asset (IASB, 2009: paras. 117-119). The available discussion paper suggests that the itemised accounting approach will be preserved. If the future standard introduces mismatches once again can only be assessed as soon as further details regarding the liability and the grant are outlined.

3.3.4 Concluding Remarks

This normative analysis of financial accounting alternatives for emission rights gave an overview of IFRIC 3 – the withdrawn treatment originally intended to be applicable for all IFRS firms – and of USofA – the rule set developed for US GAAP firm. Due to the ongoing regulatory debate, one prominent and three other proposals were additionally outlined. It was found that USofA is at least partly suitable in contracting situations, whereas IFRIC 3 largely fails in assuming any of the financial accounting functions. It can be predicted that the proposed full fair value amendment can mitigate some of these shortcomings and can furnish finer information.

The arguments presented in the recent IASB discussion paper show that the forthcoming standard for emission trading schemes is envisaged to disclose the gross economic impact of the regulation, as it was already the approach of IFRIC 3. In doing so, the Boards should carefully avoid introducing inconsistent valuation principles once again. For example, the proposal of measuring assets and liabilities on a joint basis may be appropriate to circumvent accounting mismatches, but equally violates the conception of assets and liabilities as proxies for future cash flows. Additionally, if allowance and agriculture grants will be treated analogously (and hence be measured at fair value), this should ex ante exclude deliberations on historical costs for granted assets, giving rise for divergent bases for valuation.

While it is hard to estimate the value of accounting information for contracting purposes empirically, predicting its usefulness for investors is more straightforward as their reactions are settled on the capital market. For this reason, Section 4.7 employs a regression model to assess the degree to which the information function of the four major treatments is fulfilled.

3.4 Some Resulting Question for Empirical Research

Chapter 3 presented a normative analysis towards competing approaches of how to map emission rights in management and financial accounting. Imputed costs were found to yield superior information than accrual costs with respect to both functions of performance reporting. A mixed picture arises for financial reporting: all of the major treatments face problems, most notably netting (USofA), mismatches (IFRIC 3) or potential volatility risks (proposed amendment to IAS 38). The earnings impact under the former approach seems to be more suitable in contracting situations, and the balance sheet items of the latter two treatments in an equity valuation setting. It is apparent that the related normative accounting literature cannot sufficiently predict the usefulness of accounting treatments.

Apart from illustrating that e. g. mismatches distort the efficient processing of information, this type of analysis remains silent on the question how related parties like investors, creditors or board members react to either of the accounting treatments.

Even though prior empirical research indicates that non-financial measures seem to give an idea of how firms are affected by emission trading, there is a lack of empirical insight on how the information conveyed by the available reporting alternatives is valued. Consequently, it can be asked which accounting treatment gives rise to financial reporting items that are priced according to their economic function.

Two research approaches are available to shed light on this issue: On the one hand, expectations and decisions of economic agents can be illustrated by means of complex models (see only Göx and Wagenhofer, 2009). This implies strong assumptions towards the underlying economic variables, but can also yield numerical results for unobservable decision situations. On the other hand, an empirical assessment of accounting information aims at identifying whether the reactions of subjects under scrutiny are in line with expectations. Converse implications can be found: many of the restrictive hypotheses can be relaxed, but only publicly observable information can serve as input. The following analysis investigating into the effects of accounting for emission rights will rely on both approaches. A simple model is used to simulate the ETS-specific accounting outcome, serving as input to an empirical assessment of investor expectations.

Section 4.7 will shed light on how the four major accounting treatments for emission rights process information in a capital market valuation setting. To this end, the pricing of modelled ETS-related earnings are analysed by means of value relevance regressions separately estimated for either of the alternatives. Finding pricing patterns consistent with rational expectations informs whether financial accounting information satisfies its information function. A further analysis of management accounting for emission rights is not conducted in this study since data regarding firm-level implications is not readily accessible and needs to be collected via surveys. Its availability is essential for both, purely empirical research and calibrating an analytical model.

4 Capital Market Implications of Trading in and Accounting for Emission Rights

4.1 Development of Research Questions

The previously derived questions for empirical research address important research gaps concerning the environmental and energy policies as well as accounting for emission rights. This section is to pick up these questions and to sketch possible outcomes of the subsequent empirical investigations. In the remainder of Chapter 4, the models used in this study are described in a general way (in Section 4.2), the adapted equations to answer the specific questions are presented (in Section 4.3), and finally evidence for the raised questions is presented (in Sections 4.5 through 4.7, respectively).

Considering the underlying economic mechanisms of emission trading schemes as well as the legal framework as adopted in the EU, a first research question can be formulated as follows:

> **Q1:** Does the EU ETS introduce distributional effects between the regulated energy firms and other market participants?

The concurrent theories on the effectiveness of public policy allow for three possible outcomes when empirically answering this question. Firstly, consistent with the public interest hypothesis that a regulation distributes wealth from the firms to the consumers, it can be assumed that a trade in emission rights has a negative impact on the capital market value of listed firms. The reason is that curbing pollution by rights, emission affects the production of regulated firms and is likely to lead to additional costs which cannot be completely passed on to customers. An ETS sets out to reduce the polluting production, to purchase additional certificates covering excessive emissions or to invest in abatement measures. This effect is said to hold even in case of grandfathering, since the usage of free certificates causes opportunity costs. Accordingly, there is evidence that the emission trading schemes in the USA led to an adverse capital market effect for coal mines and power producers (Kahn and Knittel, 2003: 22-23; Linn, 2006: 14-15).

Secondly, in line with capture theory, it is possible to find positive economic consequences for some groups. This would show if regulated firms are not only able to pass on the additional costs arising from the EU ETS to their customers but if imposed regulatory costs will even be over-compensated by the buyers. If this were the case, regulated firms would generate additional income from their customers with reference to the cost of abatements. In all industries that are investment-intensive, barriers to entry account for a significant stake in income generated at the expense of the customers, as capital intensity and economies of scale restrict the access of new competitors (Bain, 1954: 20; 35-36). Following industrial economics, a high bargaining power and a high concentration (Porter, 1991: 108; Makhija, 2003: 437) can result in benefits from rising factor costs. Consequently, structures as found in the European power industry probably cause rents arising from emissions trade. These arguments can also be found in recent papers employing either a Cournot or an equilibrium model calibrated with empirical data. There, it is simulated that firms are likely to benefit from the trading scheme by cutting production and simultaneously increasing earnings (Sijm et al., 2006: 59, 62; Smale et al., 2006: 42; Demailly and Quirion, 2008: 2015). Increments to earnings may also be expected when allowances are grandfathered. Only substantial emission targets which necessitate technical and economic abatement measures are likely to lower these effects (Goulder and Parry, 2008: 164-165). Abadie and Chamorro (2008: 3007) show that investing in a carbon capture and storage facility maximises future cash flows only if allowance prices exceed 55 EUR/t. The current legal framework with a generous cap and low certificate prices thus favours a business as usual emission strategy and potentially gives rise to regulatory rents.

Thirdly, it can be assumed that emission trading has no economic consequences at the firm level. One rationale behind this is that neither of the theories on the effectiveness of public policy holds. In this case, the entities may be able to pass on any additional burden arising from the scheme to their customers. A firm can choose a strategy that allows attaining the regulatory emission target at no additional cost. Another reason for insignificance might be due to emission market effects: as presented above, the spot price for allowances plummeted from almost 30 EUR/t in 2006-04 to around 0.03 EUR/t at the end of the first trading period in early 2008. Hence a business-as-usual strategy is not going to cause a significantly higher outlay. Neither attributes to economic insignificance will cause a negative effect on corporate net income nor will they affect economic decision making.

To answer Q1, the capital market effect of the regulation is estimated empirically. A correlation between the price for allowances and expected future profits will indicate investors' expectations towards emission rights: if capital market participants expect decreasing profits of regulated firms due to a rising price for emission rights, certificates and shares will behave in an inversely proportional manner. If investors consider a positive correlation between rising quotations of pollution rights and undisclosed rents, a higher price for allowances will then contribute to profits and will lead to a more favourable evaluation of the affected firm's common stock. Finally, if shareholders do not account for the price of pollution allowances in their valuation, a change in market value of regulated firms cannot be attributed to respective certificate price movements.

The second research question draws on the interaction of energy market regimes and the EU ETS. It asks:

Q2: Do energy market regimes determine the extent of distributional effects introduced by an ETS?

This question addresses the fact that the underlying market regime can be linked to an energy firm's possibility to e. g. over-compensate for the regulatory burden. It asks in which way the mode of economic interaction between the regulated firms and other agents is accomplished: by means of market mechanisms, by means of state intervention or by means of corporatist arrangements. Consequently, this perspective abstracts from two common assumptions of regulatory economics. One of these intuitions posits that the relative strength of a regulator vis-à-vis a concentrated industry structure can explain economic rents (as addressed in Q1). A second implication of the capture theory is that regulation itself is beneficial for the firms subject to it since it "buffers" the effect of external shocks on income (Peltzman, 1976: 230). If only the possibility of new entrants and selling prices are regulated, a large amount of volatility inherent in revenue streams is eliminated. Basing research on the market design, however, relaxes to control for the extent of public regulation. This is important in cases when the regulatory intervention is not observable, for instance when the state is among the firm's shareholders or when a community adopts rules by means of centralised negotiation.

It is not straightforward to hypothesise how a respective market regime influences the possible rent distribution. In a perfectly competitive market absent any regulatory intervention, the forces of demand and supply patterns determine the extent to which a firm is able to over-compensate for emission costs. In such a market with a given output, the producers will be able to introduce ETS-related costs if the demand is high and will have to incur the regulatory burden if the demand is low. This trivial observation has special implications to the production of electricity due to its special constraints. Since the demand changes constantly but the output cannot be stored, windfall profits will occur in peak hours. Additionally, the price for power will be affected by a cost pass-through when fuels are employed inducing regulated emissions (Sijm et al., 2006: 58-59).

A state-based market regime allows for a more clear-cut assumption. This arrangement cannot be compared with the institutional setting of a competitive but regulated market as found in most regulatory economics examples (e. g. Stigler, 1971). Here is no need for an external regulatory authority since the public supervision of the firm is accomplished by means of board membership. If maximising the aggregate welfare is in the government's scope, it can still be assumed that the SOE will internalise the externality of pollution, and hence bear the regulatory costs. In communitarian regimes, strong private institutions interact with the market mechanism and with public forces (Streeck and Schmitter, 1985: 128). The corporatist power market hence relies on a self-regulation in regional monopoly- or oligopoly-like arrangements. In a suchlike approach with a major participation of corporations, it can be predicted that no internalisation of the ETS will take place. Additionally, since this design does not imply a wholly unregulated market either, less efficiency than in the liberal setting is to be expected.

Prior literature only gives empirical results for the case of generally weakly versus generally strongly regulated markets (e. g. Clarke, 1980; Norton, 1985). It fails in isolating the effect of distinct state intervention modes and neglects differentiating in underlying market regimes. To answer Q2, again the same capital market model as above is employed. This estimation does not directly assess the realised profits but yields the association between the observed allowance price changes and the investor expectation towards a firm's share price. Evidence on energy market settings determining the extent of ETS-related profits or losses is given by positive or negative correlations.

The third and final empirical question addresses the mapping of an emission trading scheme to accounting. As set out, economic realities are transformed to accounting numbers in several alternative ways. Apart from the normative analysis presented above, it is possible to estimate their competing properties in a market valuation setting. In doing so, this question is formulated as follows:

Q3: Which accounting treatment gives rise to financial reporting items that are priced according to their economic function?

All financial reporting alternatives have relative merits and shortcomings: IFRIC 3 offers comprehensive disclosures of balance sheet items but gives rise to a potential measurement or reporting mismatch in accounting income (Günther, 2006: 234), which results from different treatments of the asset, the government grant, and the liability. The proposed IAS 38 amendment also offers full disclosure but suffers from the potential shortcomings of a full fair value treatment such as volatility in the income statement. USofA reduces the mismatch possibilities but only discloses the net effect of the emission regime. Consequently, it can be expected that full fair value models measure assets, liabilities and income in a manner relevant in investment decisions (Barth, 1994: 12). The mixed models of IFRIC 3 are likely to inherit the convenient properties of balance sheet items measured at fair value but to lack in conveying meaningful earnings. Finally, netting potentially foregoes an informative balance sheet for the sake of relevant accounting income.

To answer Q3, it is tested whether assets and revenues (liabilities and expenses) are positively (negatively) correlated to market values. This pinpoints a correspondence with the underlying economic function of the financial reporting items, and this can be referred to as consistent valuation. Additionally to this question it can be considered if one accounting treatment dominates the others. This is the case if it has more financial reporting items which are consistently valued than another one and none less. Prior literature neither provides empirical evidence on whether the respective accounting treatments lead to a disclosure of economically meaningful balance sheet and income statement components nor whether one of the methods seems to be superior to others. This paper tries to shed more light on these questions by comparing the accounting alternatives' respective properties in a market valuation setting.

4.2 Models in Empirical Capital Market Research

4.2.1 Detecting Pricing Effects

Pricing effects arise in a capital market setting when investors assign a specific value to a designated security. Estimating these pricing effects can be accomplished by means of the Capital Asset Pricing Model (CAPM) that measures how strongly a share reacts to overall market movements. Underlying intuition is that in efficient capital markets any new relevant (publicly available) information is incorporated in the market portfolio (Fama, 1991: 1575). Hence, investors consider this news concerning purchasing or selling securities of listed firms which makes this information price-relevant (already Ball and Brown, 1968: 163). Following the seminal articles of Sharpe (1963: 281), Lintner (1965: 601) and Mossin (1966: 778), estimating the investors' expectation towards the value of a certain security has been the basis to a large strand of empirical literature. The general form of the model employed in this literature can be written as follows:

$$R_{it} = \alpha + \beta R_{mt} + \varepsilon$$

where:

R_{it} = common stock return of firm i in period t;

R_{mt} = return of the market portfolio in period t.

In this regression equation, the intercept termed Jenson's (1968: 394) α indicates excess (inferior) returns of a share taking a positive (negative) sign. Consequently, if the capital market analysed is in equilibrium, a constant of zero is expected (Binkley et al., 1996: 26). The slope coefficient β captures the systematic risk that is the sensitivity of the security towards the market. A β above (below) one indicates a share yielding greater (smaller) returns than the underlying portfolio (Schmidt and Terberger, 2003: 358). Formal model derivations usually define the firm-specific and market stock returns as netted by the rate of return on a risk-free investment. In empirical research, this return can be proxied by the interest rate on government bonds (Fama and French, 1997: 155), but is also often omitted (de Jong et al., 2006: 8-9; Ziegler et al., 2007: 360).

It should be noted that the standard CAPM relies on restrictive assumptions, like the presence of complete capital markets, and infinite divisibility of assets (see for an overview Jensen, 1968: 86). Inevitably, some of these assumptions will be (at least partially) violated in empirical specifications. Consequently, this market model can possibly suffer from a non-stationarity in the time series of β coefficients (Black et al., 1972: 15) and the general limitation of introducing only one single source of information – market returns. When additional factors influence the value of a share, a multifactor market model can be set up with theoretical bases like the Intertemporal CAPM (Merton, 1973: 884-885) or the arbitrage pricing theory (Ross, 1976: 355). Both include other factors in addition to the aggregate market driving stock prices (Fama and French, 2004: 38); their generalised augmented equation reads as follows:

$$R_{it} = \alpha + \beta R_{mt} + \sum_{j=1}^{k} X_{jt} + \varepsilon$$

where:

X_{jt} = vector of k additional explanatory variables observable in period t.

The intuition is that in a multifactor market model, various variables with relevance for stock price returns can be tested. This includes macroeconomic indicators as e. g. changes in inflation rates or manufacturing output (Vernimmen, 2005: 428). Models in this vein have been applied in studies on the gold price as risk factors for the mining (Blose and Shieh, 1995: 136; Tufano, 1998: 1032-1033; Faff and Chan, 1998: 25; Faff and Hillier, 2004: 480-481) and the oil price for the extracting industries (Faff and Brailsford, 1999: 79-80; Sadorsky, 2001: 24-25; El-Sharif et al., 2005: 825, 828), as well as in studies on exchange rates being a relevant risk factor for import and/or export oriented firms (Jorion, 1990: 337; He and Ng, 1998: 739; Nydahl, 1999: 254; Allayannis and Ofek, 2001: 280; Griffin and Stulz, 2001: 229-230; de Jong et al., 2006: 23-24; Muller and Verschoor, 2006: 204).

In answering the research questions Q1 and Q2, a multifactor model will be used since the simple CAPM fails in the case of an ETS as two markets impose an unknown pattern on the stock price. To explicitly control for changes in allowance prices, it is necessary to add this information as an additional explanatory variable. This allows using the capital markets as a gauge for distribution: neither the cost nor the revenue component are analysed separately, but the net pricing effects. This approach is superior to examining, for example, solely product prices since introduction of the ETS. Further, investigating into the pricing effects of different energy market regimes follows the early contribution of Peltzman (1976: 230). Drawing on single factor models it was shown that regulatory interventions can induce shocks (Sawkins, 1996; Dnes and Seaton, 1999) or persistent changes (Clarke, 1980: 351-353; Norton, 1985: 680-681) and lower the systematic risk. Despite the general deficiencies of the CAPM and the criticism regarding its suitability in testing for regulatory influence (Chen and Merville, 1986; Ahn and Thompson, 1989), it remains the major model for market-based research in this domain. Considering that studies with adjusted market models largely support the previous findings of this strand of literature (Riddick, 1992; Davidson et al., 1997; Binder and Norton, 1999), additional tests regarding the pricing patterns of the pre-EU ETS era will also rely on the simple CAPM.

4.2.2 Detecting the Relative Usefulness of Accounting Information

Usefulness of information is the central property of a financial accounting system used in equity valuation. Its information is regarded as useful if it conveys relevant and reliable information in a decision making process, and draws on the observation that market participants base investment decisions on accounts which allow forecasting the future performance of a specific firm. In the empirical accounting literature, a well established strand of research uses so called value relevance models to assess this information content. There, value relevance is hypothesised to take effect if accounting disclosures are associated with equity market values or changes in it (Barth et al., 2001: 79). An analysis like this juxtaposes stock prices as proxy for the present value of future net cash flows and the accounting value of separable net assets (Kothari and Shanken, 2003: 74). In this domain, contributions usually rely on two alternative designs: event and association studies. The event study approach asks whether a specific accounting announcement causes market reactions such as increasing trading volumes or direct pricing effects over a short time period around a specific event (Kothari, 2001: 116). If relevant information disclosed in a specific announcement was (at least to some extent) not known to the market before, a corresponding market reaction is assumed. Studies like Ball and Brown (1968), Beaver (1968) or Amir and Lev (1996) test for a timely market information processing to a specific value relevant event. The association study design inquires into the correlation of accounting and market measures without hypothesising a causal relationship (Kothari, 2001: 116). This is usually accomplished in regression models explaining market performance by means of financial variables, estimated over longer, contemporaneous time periods like one quarter or one year. Following this methodology, estimates for value relevance are the regression's explanatory power and the slope – the so called earnings response coefficient.

Association studies can either be specified as price or return models. Generalised versions of these models can be expressed as follows:

$$P_{it} = \varphi + \omega_1 EARN_{it} + \omega_2 BVE_{it} + \varepsilon$$

$$R^{\delta}_{it} = \vartheta + \upsilon_1 EARN_{it} + \upsilon_2 EARNCH_{it} + \varepsilon$$

where:

P_{it}	= price of the common stock of firm *i* in period *t*;
$EARN_{it}$	= earnings of firm *i* in period *t*;
BVE_{it}	= book value of equity of firm *i* in period *t*;
R^{δ}_{it}	= discrete common stock return of firm *i* in period *t*;
$EARNCH_{it}$	= change in earnings with respect to the same period of a year earlier of firm *i* in period *t*.

In the first case, market prices are regressed on balance sheet numbers, in the second case, returns, as changes in market values, are regressed on changes in balance sheet numbers such as earnings and/or earnings changes. The former, the so called price specification, is rooted in the Ohlson (1995) model, in which the market value of a firm is derived from accounting value of equity and earnings. Previous studies like e. g. Collins et al. (1997), Core et al. (2003) or Barth et al. (2008) employed price specifications in estimating value relevance. The latter model, usually referred to as return specification, can be obtained from the price specification by taking the first difference of the variables on both sides of the regression equation. Examples in literature uniquely employing this or similar models include Easton and Harris (1991), Alford et al. (1993) and Bartov et al. (2001).

It has, however, to be noted that there are also some concerns against value relevance as a measure for accounting quality. As e. g. found in Francis and

Schipper (1999: 336-340) or Ryan and Zarowin (2003: 353-358), measures for value relevance exhibit instable properties over time. This effect is attributed to correlated omitted variables with varying time characteristics (Kothari and Shanken, 2003: 78) as well as to timing differences between cash flows and accounting earnings (Holthausen and Watts, 2001, 2001: 58). Apart from that, heteroskedasticity, measurement errors and cross-sectional differences in valuation parameters can induce a statistical bias in estimators. Some of these issues can be easily controlled for: using White (1980) adjusted standard errors mitigates the statistical effect of inflated *t*-tests due to heteroskedasticity, and a cautious scaling reduces the omitted variable problem of firm size (Brown et al., 1999; Easton and Sommers, 2003). Since these shortcomings apply to price models in a more prominent manner, many authors employ both price and return studies in order to provide for robust results (e. g. Kothari and Zimmerman, 1995; Francis and Schipper, 1999; Harris et al., 1994; Barth et al., 2009).

In a retrospective, value relevance researchers were the first to establish empirical methods in accounting (Ball and Brown, 1968; Beaver, 1968), and studies following this design dominated the literature during the 1990s. At the beginning of the subsequent decade, several authors voiced criticism over its methodological shortcomings and directed the focus to other open questions like conservatism or earnings management. Still, there seems to be a revival of value relevance research studies recently, as employed in Cahan et al. (2009), Jenkins et al. (2009), Kirkulak and Balsari (2009), Kousenidis et al. (2009) or Kang and Zhao (2010). In answering research question Q3, this study contributes to the case of accounting information for emission rights with the help of value relevance models.

4.3 Specification of the Models Employed

4.3.1 Pricing Effects of Emission Rights

Following the arbitrage pricing theory (Ross, 1976: 355), a multifactor market model is specified to capture the incremental economic influence caused by the new regulatory scheme. This allows for a separated test whether the emission trading leads to changes in a firm's wealth. More precisely, a firm's exposure towards an ETS can be estimated as follows:

$$R_{it}^{ld} = \alpha + \beta_1 R_{mt}^{ld} + \beta_2 \, {}^{X}R_{et}^{ld} + \varepsilon \qquad (1)$$

where:

R_{it}^{ld} = logged daily return on the common stock of firm i in period t;

R_{mt}^{ld} = logged daily return on the market portfolio in period t;

${}^{X}R_{et}^{ld}$ = logged daily return on emission allowances in period t;

X = $X = S$ denotes spot contracts, and $X = F$ denotes futures contracts which mature in 2008-12.

In this equation, α denotes the intercept, ε the error term, and the coefficients β_1 and β_2 capture the systematic risk as well as the exposure towards the trade in pollution rights. In order to mitigate the effect of heteroskedasticity, model (1) is estimated using White (1980) adjusted standard errors. Analogous to a single factor β, the second slope coefficient quantifies the change in returns on common stock of a firm at a 1% change in returns on emission certificates. In order to answer research question Q1, whether the ETS gives rise to rents between the regulated energy firms and other market participants, special attention is paid to the statistical properties of β_2. If the regulation is perceived by capital markets as an additional relevant factor for valuation, the coefficient on the indicator variable in the market model should be significantly different from zero. In case of a negative coefficient, higher returns on the market for emission rights lead to higher (opportunity) costs at the firm level, which is bad news when firms are not able to shift the additional burden to stakeholders, particular customers. Then the trading scheme should have a negative effect on firm value, since the dischargers carry the regulatory burden. Finding a positive coefficient on the indicator variable is consistent with regulatory spill-overs into product markets.

The economic impact of the EU ETS on the industry may depend on the rigour of the regulatory environment. While the first trading period from 2005-03 until 2008-03 was seen as an implementation and testing period, the EU Member States agreed on meeting the Kyoto targets during the second period ending 2012. That is why the spot market for emission certificates may only be of legal relevance since it is the place where obligations for the actual period are traded. It may not be relevant for economic decision making on avoiding emissions, or it may give rise to regulatory rents. No prediction as to the sign of the beta coefficient on the spot contracts β_2 is offered since it is ex ante unclear what kind of distributional effects are introduced by the regulation, if any. To take into account the effects of the allowance market collapse in 2006-04, it is necessary to estimate the model over several sub-periods. It can be supposed that for periods after the sharp decline in allowance prices the ETS coefficient will not yield any significant results.

The futures markets for emission rights also play an important role from an economic perspective. It has been shown that certificate contracts for the running and for the upcoming period behave differently (Trück et al., 2006: 23). Hence, it is likely that investors expect a continuing regulatory tightening. Taking a similar but probably more persistent anticipation of ETS-related distributional effects into account, it is predict that the coefficient of the futures contracts will take the same sign as the equivalent for spot contracts. This is hypothesised to apply for all sub-period betas, regardless of demand and supply issues of the current trading period.

4.3.2 Interaction of Energy and Allowance Markets

The multifactor market model as specified in equation (1) is also employed when researching into the interacting effects of European energy and emission markets. All sampled energy firms are attributed to either the market-, state- or community-based regime according to the country of incorporation. The empirical analysis is repeated for each of the three subsets in order to answer question Q2 – whether the market regime determines the extent of distributional effects due to an ETS. If different estimates for the allowance pricing coefficient β_2 can be found, the research question can be confirmed. It is predicted that state-based regimes hamper distributional ETS effects in favour of the producers, at least in the short run: if public influence on these firms is exerted via board memberships, it can be assumed that it is business policy to peg selling prices for some consumer groups. A quick over-compensation of emission (opportunity) costs is hence less likely. No prediction can be given for the other market regimes, since neither the maturity of the European market-based regimes nor the properties of corporatist regimes have yet been assessed from a capital market perspective.

The proposed design bears the econometric drawback that the coefficients in question cannot be easily tested for statistically significant differences since regressions are estimated for separate populations. An alternative is to specify dummy variables for the different regimes and to interact them with explanatory variables. Then, an interacted parameter different from zero indicates a significant in- or decrement to the basal coefficient. These models are statistically equivalent to separate estimations, but less straightforward to interpret. For this reason, the more intuitive approach is tabulated while unreported results can additionally furnish a test on significance.

4.3.3 Usefulness of Accounting Information on Emission Rights

An assessment of the accounting information on emission allowances is performed for the following four treatments: the cost method in IFRIC 3 (hereafter abbreviated as COST), the revaluation model in IFRIC 3 (termed REV), the proposed full fair value amendment to IAS 38 (labelled FFV), and the approach under USofA 101 (referred to as NET). In doing so, both the price as well as the return specification is used: this is particularly important in the given setting not only in order to provide for robust results, but also to assess the consequences for earnings and balance sheet numbers. Taking advantage of the component-based approach as first developed in Barth (1994), net equity (earnings) is adjusted in order to separate the effect related to the accounting treatment under scrutiny.

In component-based approaches, adjustments are usually performed using real data obtained from the respective financial statements or the notes (e. g. Barth et al., 1992). However, such data is only rarely available in the emission allowance case. With respect to the available sample, the respective effects on the balance sheet and income statement are disaggregated for only one fifth of the firm quarter observations. This implies that there is insufficient data available to compare the ETS-related assets, liabilities and income statement components across accounting treatments. Thus, consequences of the respective alternatives are modelled based on the following three assumptions:

1. The physical emission assumed to be proxied by quarterly firm-specific revenues. To exemplify this assumption, consider an entity's annual revenues are distributed over the fiscal year as follows: 25% in the 1st Quarter,

20% in the 2nd Quarter, 25% in the 3rd Quarter and 30% in the 4th Quarter. If its verified emissions add up to 1,200,000 t on 2005-12-12, the quarterly emissions are proxied as being equivalent to sales, or: 300,000 t in the 1st Quarter, 240,000 t in the 2nd Quarter, 300,000 t in the 3rd Quarter and 360,000 t in the 4th Quarter.

2. Following the legal framework, granted allowances are supposed to enter the balance sheet on every March 1, the first day of the trade, and to be surrendered on every April 30. To emulate an emission right market transactions pattern, the excess (shortfall) of certificates is sold (purchased) in tranches according to quarterly emissions. Assume the same entity as before was allocated only 1,000,000 certificates on 2005-03-01, and is hence exposed to a shortfall of 200,000 units. The purchase of the lacking allowances is proxied to take place at the end of each quarter according to emissions, or: 50,000 units in the 1st Quarter, 40,000 certificates in the 2nd Quarter, 50,000 rights in the 3rd Quarter and 60,000 allowances in the 4th Quarter. There is no further speculation.
3. The allowances are assumed to be used to settle the obligation incurred during the twelve months starting on January 1. If an entity did not fulfil its obligation to surrender sufficient emission rights in a trading year, this shortfall is carried forward to the consecutive year.

First a price model is specified to assess the information content of balance sheet numbers related to ETS accounting. To capture the effect of the balance sheet item of interest, that is the specific asset or liability emerging through application of the respective ETS accounting treatment,[38] the model of Aboody and Lev (1998: 172) is adapted accordingly. Due to high correlations between the assets and liabilities in each approach, the items need to be tested in two separate regressions to ensure unbiased estimators using the following price regression:

$$P_{it}^{q} = \varphi + \omega_1 EARN_{it}^{q} + \omega_2 {}^{adj}BVE_{it}^{q} + \lambda ETS\_BSI_{it}^{q} + \varepsilon \quad (2)$$

where:

P_{it}^{q} = price of the common stock ending one month after fiscal quar-

[38] The valuation effect of a grant possibly emerging remains untested as this item only represents deferred income that should not have a valuation effect.

ter end of firm i in period t, scaled by the price at fiscal quarter beginning;

$EARN^q_{it}$ = earnings before extraordinary items of firm i in quarter t, scaled by the market value at fiscal quarter beginning;

${}^{adj}BVE^q_{it}$ = book value of equity, adjusted for the ETS balance sheet item (either allowances asset or emission liability), scaled by the market value at fiscal quarter beginning, of firm i in quarter t;

$ETS\_BSI^q_{it}$ = ETS-specific balance sheet item (either allowances asset or emission liability) according to one of the four alternative treatments of firm i in period t, scaled by the market value at fiscal quarter beginning.

Following Brown et al. (1999) and Arce and Mora (2002), all accounting variables are scaled by the opening market value since this is the most promising approach to deal with the econometric shortcomings of price regressions. Further, all modelled variables are winsorised at the 1% and 99% percentiles. To control for heteroskedasticity when estimating regression (2), *t*-tests are calculated using White (1980) adjusted standard errors. The models also include quarter fixed effects as the time series of net profits in interim reports might be influenced by seasonal effects, like the turnover in power, heat, or cement potentially driving net profits. Quarterly reports may not show similar characteristics on interim date, as some are audited and others are not (Basu et al., 2001: 4). This also implies a necessity to control for quarter fixed effects. Finally, industry fixed effects are added even though all of the four examined industries share some similar characteristics like high market concentration and strong barriers to entry. A control for country accounting effects appears not to be necessary since all EU-based listed firms are required to adopt IFRS as the only basis to their consolidated reports (there is a temporary opt-out for previous US GAAP adopters; it does not play an important role in the selected sample). Further distinguishing between energy market regimes is inconsistent since the empirical analysis focuses on the aggregate usefulness of accounting information and for that reason also includes non-energy firms.

As each of the four treatments can give rise to both an asset and a liability, $ETS\_BSI^q_{it}$ can theoretically take eight different values. However, it is sufficient to estimate four different versions of model (2) due to similarities and modelled simplifications in the accounting treatments: the asset is equal for the REV and FFV models, the liability does not differ in the COST, REV and FFV models, and due to the modelling assumptions a possibly existing NET asset is amortised fully and cannot be considered. It is then looked at the coefficients λ of the balance sheet items in the different versions of the model to check whether the considered items have the expected signs, shedding light on the third research question Q3, which of the financial reporting items are valued in accordance with their economic function? Coefficients on $ETS\_BSI_{it}^{q}$ should have positive signs for assets and negative signs for liabilities. To further find an accounting treatment dominating the others, a consistent pricing of assets and liabilities is additionally required.

The component approach in returns models disaggregates income figures with regard to the particular effect attributable to the accounting treatment under scrutiny; in this case, the earnings effects of emission rights need to be separated. In line with the specifications used in the prior literature (e. g. Lev and Sougiannis, 1996; Aboody and Lev, 1998), the following model is estimated:

$$R_{it}^{\delta q} = \theta + \upsilon_1 \, {}^{adj}EARN_{it}^{q} + \upsilon_2 EARNCH_{it}^{q} + \kappa ETS\_EARN_{it}^{q} + \varepsilon \qquad (3)$$

where:

$R_{it}^{\delta q}$ = discrete buy-and-hold common stock return for a three-month period ending one month after fiscal quarter end of firm *i* in period *t*;

${}^{adj}EARN_{it}^{q}$ = adjusted earnings before extraordinary items (i. e. reported income plus/minus ETS earnings impact according to one of the four alternative treatments), scaled by the market value at fiscal quarter beginning, of firm *i* in quarter *t* (i. e. $EARN_{it}$ - ETS_EARN_{it});

$EARNCH_{it}^{q}$ = change in quarterly earnings before extraordinary items with respect to the same quarter of a year earlier, scaled by the market value at fiscal quarter beginning, of firm *i* in period *t*;

$ETS\_EARN_{it}^{q}$ = quarterly net ETS earnings impact according to one of the four alternative treatments, scaled by the market value at fiscal quarter beginning, of firm *i* in period *t*.

As done in Aboody and Lev (1998: 171), all accounting variables are scaled by the market value at fiscal quarter beginning; additionally, the modelled variables are winsorised at the 1% and 99% percentiles. All regressions are performed using White (1980) standard errors and include quarter and industry fixed effects.

Four estimations of model (3) are performed in which $ETS\_EARN_{it}^{q}$ captures the income statement consequences of the COST, REV, FFV and NET models, respectively. Regression results are interpreted analogously to the price regressions. The signs of the coefficients on $ETS\_EARN_{it}^{q}$ in the four estimations reveal whether earnings supply good and losses bad news and help answering Q3.

4.4 Data and Sampling

Empirically answering the research questions requires the definition of two different samples: sample S1 comprising a representative set of electricity firms to detect pricing properties distinctly associated with the energy markets (Q1, Q2), and sample S2 comprising a representative set of (energy and non-energy) entities with sufficient available accounting data to detect the aggregate effect of ETS-related accounting information (Q3). Using the same sample for all of the questions is inefficient: a broader sample obviates a focussed analysis of the capital market implications to power generation, a smaller sample (restricted to the energy market) does not allow for general conclusions as to accounting. As Figure 7 shows, the samples overlap for those firms meeting the both sample selection requirements. The detailed procedures are outlined in the remainder of this section.

The assessment of ETS-induced distributional effects by means of a capital market model necessitates selecting listed firms only. Obtaining the stock price data from *Datastream*, firms in the *Electricity* and in the *Gas, Water & Multi-utilities* categories were included. This gross sample comprising 128 EU-based entities needs to be adjusted for all those firms not subject to the regulation: out of them, 12 generate electricity without any regulated emissions; 6 firms were identified as equity holdings; another 15 only transmit, distribute or supply electricity; 29 entities are water, waste or gas utilities with no own power

Figure 7: Graphical Illustration of Samples Used

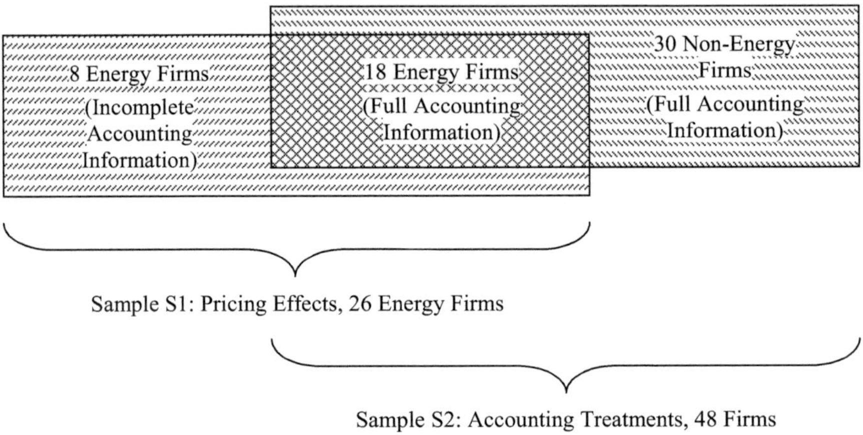

Table 15: Sample Selection

Panel A: Selection of Sample S1 (Pricing Effects)

Description	Firms
Total number of firms in the *Electricity* and in the *Gas, Water & Multiutilities* of *Datastream*	128
Deleted from sample:	
Power firms without emissions of regulated gases	-12
Equity holdings	-6
Transmission, distribution or supply of electricity	-15
Water, waste or gas utility	-29
Manufacture of equipment	-16
All generation sites outside the EU ETS	-1
Matching loss	-23
Final sample	26

Panel B: Selection of Sample S2 (Accounting Treatments)

Largest installations	159
Deleted from sample:	
Not EU-based	-19
Private-/state- owned	-73
Missing quarterly data	-16
Other primary activity	-3
Further installations controlled by sample firms	
Final sample	48

generation; and 16 firms are associated with the manufacture of technical equipment; one entity only holds generation facilities outside the EU ETS. Additionally, another 23 entities had to be eliminated arising from a loss on matching between the capital market data and allocation and compliance data as published by the European Commission in the CITL. The final sample S1 contains a total of 26 firms; its derivation is summarised in detail in Panel A of Table 15. Out of the largest 500 emitting installations subject to the EU ETS, 205 ones are held only by the selected power producers. Considering the amount of verified emissions in 2005 and 2006 throughout the EU-25, these installations account for 36.94%.

The selection of sample S2 for evaluating the available accounting treatments for emission rights began with indentifying the installations of all energy and non-energy firms subject to the EU ETS as obtained from the CITL. Emitters with data missing in the database during 2005 to 2007 are excluded. In order to facilitate the matching between installations and the respective controlling entities, the 500 largest facilities were selected, held by 159 firms. Installations of 19 not EU-based parents were removed from the sample; 73 private or state-owned entities were excluded from the sample as they are not publicly quoted; another 16 corporations were dropped due to missing quarterly financial reports. Additionally, the sample is constrained to entities with a primary business in one of the regulated activities leading to the exclusion of three firms with other main activities. The final sample S2 comprises 48 listed firms with quarterly financial reports available. Expressed in terms of the ETS, this set consists of 1,001 installations or 52.23% of the aggregate verified pollution in the EU-25. Panel B in Table 15 summarises this sample selection procedure.

Table 16 illustrates the distribution of sample firms over EU Member States and (if applicable) industries. Sample S1 focuses on 26 power producers, out of which five ones are domiciled in Germany and the UK, respectively. Considering the mere count of entities cannot be ex ante considered as a sign of competitive energy markets. It rather indicates that in both named countries coal still plays a large role in electricity generation, inflating the number of ETS regulated producers. With respect to sample S2, energy and refining are by far the most prominent sectors in terms of analysed entities; cement and steel producers constitute roughly one third of the sample. The top three countries concerning the number of regulated corporations are Italy, Spain, and Germany, accounting for half of S2. Glass and paper industries are notably absent: both industries are characterised by small firms, and even many bigger entities in these sectors are privately held and therefore not subject to mandatory IFRS reporting and disclosure.

Table 16: Country of Incorporation and Industry Affiliation of Sample Firms

Country	Sample S1	Sample S2				
	Energy	Energy	Refining	Cement	Steel	Sum S2
Austria	2	2	1		1	4
Czech Rep.	1	1	1			2
Finland	1	1	1			2
France	2		1	1		2
Germany	5	3		1	2	6
Greece	1	1	2	1		4
Hungary			1			1
Italy	4	4	3	3		10
Netherlands			1		1	2
Poland			1			1
Portugal	1	1	1	2		4
Spain	4	4	2	1		7
Sweden					1	1
Un. Kingdom	5	1	1			2
Sum	26	18	16	9	5	48

Table 17: Accounting Rules Applied for Emission Rights (Sample S2)

Treatment	Adopters	No. of Observations: Empirical ETS Income
IFRIC 3	7	31
USofA	38	87
Undisclosed	3	480
Sum	48	598

Finally, Table 17 informs about how companies of sample S2 account for emission rights. At the end of the observation period, seven firms stated to employ the IFRIC 3 treatments, 38 entities used USofA in preparing their accounts and three did not specify the method choice. The choice of accounting treatments for emission rights is independent from the standard used to prepare financial statements. Further quantitative information, like the value of allowance assets, emission liabilities and their changes is only rarely found in financial statements. The "bottom line" figure of the ETS earnings impact is reported in 31 firm-quarter observations of IFRIC 3 adopters, and in another 87 observations of USofA users. The gap in quantitative disclosure makes it necessary to model the income components with the help of the additional non-financial information allowance price, allocation and compliance status. Further, partitioning into subsamples of actual IFRIC 3 and USofA preparers in order to test the computed earnings impact separately is not possible due to small sub-sample sizes.

All estimations cover, unless otherwise stated, the full first trading period of the EU ETS as the Commission implemented a stricter emission cap for the Kyoto compliance period. The capital market models employed to answer questions Q1 and Q2 are based on 19,362 daily observations. The analysis starts on 2005-04-25 marking the first return on allowance futures, and ends on 2008-03-20 the last trading day in first phase contracts. All securities quotations were used as provided by *Datastream*. With regard to the economic value-relevant factors considered in the model, different sources were used: the market portfolio is proxied by the *World Market Index* as obtained from this database (comparable to Muller and Verschoor, 2006: 200-201). Emission rights spot prices were used from the EEX and respective futures prices from the ECX. In answering Q3, the 13 fiscal quarters between 2005-03 and 2008-03 are examined, yielding 598 firm-quarter observations. All accounting data were collected and analysed manually, whereas market data were obtained from the *Global Vantage* database. Again, prices for emission allowances are obtained from the EEX, and allocation and compliance data available in the CITL are used.

4.5 Pricing Effects of Emission Right Returns in the Power Sector[39]

4.5.1 Results

This section presents the empirical results in answer to research question Q1 and furnishes several additional analyses. In doing so, the estimates for the multifactor market model as specified in Equation (1) are presented and discussed in the remainder. Panel A of Table 18 shows the coefficients of a regression on the market portfolio and allowance spot contracts for the trading period 2005-2007. A first note has to be made with respect to the market coefficient and the overall coefficient of determination, since both values are low compared to other studies employing a similar model (for example in the oil extracting industry El-Sharif et al., 2005: 825). This effect, though, is attributable to the electricity sector. Other empirical studies illustrated that the market risk of this industry is lower than that of the total market, as the expected profits depend less on economic cycles; consequently, these results are in line with previous studies (Competition Commission, 2006: K7; Grout and Zalewska, 2006: 166).

A further look at the pollution rights exhibits a small but highly significant spot contract coefficient for the complete observation period. In order to illustrate distinct effects in shorter periods, the time frame examined has been subdivided into six-month periods respectively beginning in May and November.[40] The spot contract coefficients for the first two sub-periods are up to eight times higher than for the complete period. In late 2006-04, the market reacted to the information of an excess supply in emission rights with a severe decline in spot prices. The relevance of Phase 1 certificates seems to be lost after this incident; from 2005-05 onwards, the coefficient is not significant at conventional levels except for the last term immediately prior to the end of the period. Apart from this latter effect – that may be driven by the infinitesimal returns with prices below the 0.10 EUR/t threshold – the assumption of economic insignificance of

---

39 This section is based on parts of Veith et al. (2009). The data used here were updated to cover the full first trading phase; the results remain, however, qualitatively unchanged.

40 By intention a partition according to the calendar year has been rejected for two reasons: Firstly, the trade in emission futures started on 2005-04-22; secondly, the slump in prices for emission spot contracts took place in late April 2006.

Table 18: Pricing Effects in Interim and Total Estimation Periods

This table reports regression results of firm on market portfolio returns. Equation (1) is estimated as linear regression with White (1980) standard errors:

$$R_{it}^{ld} = \alpha + \beta_1 R_{mt}^{ld} + \beta_2\ {}^{X}R_{et}^{ld} + \varepsilon \quad (1)$$

where: R_{it}^{ld} is the logged daily return on the common stock of firm i in period t; R_{mt}^{ld} is the logged daily return on the market portfolio; superscript $X = S$ denotes spot contracts, and $X = F$ denotes one year-ahead futures contracts; ${}^{S}R_{et}^{ld}$ is the logged daily return on emission allowances spot contracts; ${}^{F}R_{et}^{ld}$ is the logged daily return on emission allowances futures contracts mature in 2008-12. Two-tailed t-statistics are given in parentheses below a coefficient; R^2 is the adjusted coefficient of determination; F is the overall goodness of fit test. *, ** and *** denote levels of significance of 0.10, 0.05 and 0.01, respectively.

Period	Constant α	R_{mt}^{ld} β_1	${}^{X}R_{et}^{ld}$ β_2	N	R^2	F
Panel A: Exposure towards Emission Allowance Spot Contracts						
2005-04-25 - 2005-10-31	0.001***	0.632***	0.024***	3,210	0.0477	64.61***
	(2.648)	(10.991)	(3.072)			
2005-11-01 - 2006-04-28	0.001***	0.515***	0.037***	3,308	0.0454	72.31***
	(3.860)	(10.644)	(4.941)			
2006-05-01 - 2006-10-31	0.000*	0.753***	0.006	3,432	0.1468	243.80***
	(1.824)	(21.463)	(1.067)			
2006-11-01 - 2007-04-30	0.000	0.744***	-0.003	3,354	0.1025	159.20***
	(0.293)	(17.759)	(-1.119)			
2007-05-01 - 2007-10-31	0.000	0.677***	0.003	3,432	0.1541	238.80***
	(1.015)	(21.835)	(1.462)			
2007-11-01 - 2008-03-20	0.000	0.657***	0.003*	2,626	0.1307	100.10***
	(-0.177)	(14.150)	(1.775)			
Total Period	0.000***	0.681***	0.003***	19,362	0.1089	643.20***
	(3.617)	(35.862)	(3.005)			
Panel B: Exposure towards Emission Allowance Futures Contracts						
2005-04-25 - 2005-10-31	0.001***	0.632***	0.028***	3,210	0.0482	64.07***
	(2.663)	(10.997)	(3.301)			
2005-11-01 - 2006-04-28	0.001***	0.521***	0.023***	3,308	0.0352	62.29***
	(3.348)	(10.705)	(3.219)			
2006-05-01 - 2006-10-31	0.000*	0.774***	0.025***	3,432	0.1490	243.70***
	(1.896)	(22.055)	(3.105)			
2006-11-01 - 2007-04-30	0.000	0.738***	0.013	3,354	0.1029	159.20***
	(0.504)	(17.403)	(1.624)			
2007-05-01 - 2007-10-31	0.000	0.670***	0.027***	3,432	0.1554	242.10***
	(0.690)	(21.729)	(2.831)			
2007-11-01 - 2008-03-20	0.000	0.599***	0.146***	2,626	0.1462	125.00***
	(-0.444)	(12.739)	(5.889)			
Total Period	0.000***	0.677***	0.029***	19,362	0.1113	662.40***
	(3.290)	(35.812)	(7.762)			

spot contracts after the market collapse can be confirmed. These results allow as a first conclusion that investors perceive a trade in emission certificates as an additional factor in valuing the regulated firms only as long it seriously impacts earnings: an ETS with certificate prices ranging marginally above zero – as was the case throughout the later sub-periods – guarantees for inexpensive business-as-usual emissions.

Supplemental to an assessment of emission spot contracts, it is necessary to analyse the predictions towards the second trading phase between 2008 and 2012. Statistically, these beliefs are captured in the slope coefficient for allowance futures maturing in 2008-12 as presented in Panel B. With regard to the complete observation period, a coefficient can be found that is again positive and significantly different from the spot contracts coefficient: performing a Chow test on inequality is accepted ($F = 44.52$, $p(F) = 0.000$). It has to be concluded that the market participants predict a higher economic impact for power producers for the first year of the new round of Kyoto trading, which began in 2008. The changes in allocation modes and a reduction of the emission target for special sectors like power generation as announced by several national governments and the European Commission are predicted to be more restrictive. Furthermore, the results for the sub-periods exhibit a picture which does not correspond to the spot contracts; here a high significance of the coefficients can be observed in all but one period.

Taking the results for the spot and futures contracts together, they do not allow for the assumption of economic insignificance due to the trade in pollution rights. Rather, the positive correlations in both estimations are the central finding that an increase in emission allowance prices is correlated to an increase in the examined firms' share prices. Investors seem to expect higher future earnings in case of positive returns on the allowance markets, and vice versa. Consistent coefficients in the base model allow estimating the valuation effects of the EU ETS. Fitting market and certificates returns into the estimated regression equations of the respective sub-periods yields the multiple of stock returns on the market portfolio as well as on allowance returns. Hence the influence of rights price changes on the estimated return of the selected firms can be quantified. During the first six trading months, the increase in certificates prices positively impacted the estimated average stock market returns by 0.6%. Afterwards, with the crash of the emission allowance market the spot market effect curbed the stock return; with respect to the complete sample period the estimated reduction amounts to 2.0%. The expectations for the second trading period are different; even though the return on futures contracts was small, investors predict a positive wealth effect in 2008 resulting in an increment of 0.7% on stock returns. These findings are in line with financial markets perceiving that the European electricity industry is not only likely to pass on a material share of the regulatory burden to customers, but that it may even achieve regulatory rents. In order to control for the robustness of the employed market model, four additional tests are performed.

4.5.2 Additional Analyses

4.5.2.1 Control Group

As a first robustness check, it seems necessary to address the question whether the ETS is able to raise the predicted income from the entire electricity industry or only from those entities involved in the regulation. Electricity generators are able to employ different technologies or fuels with distinct effects on emissions. Some entities, e. g. operating nuclear or hydro plants as well as solar or wind parks are hence entirely outside of the scheme and should not be affected. First, all generators supply the same good in the same market and are likely to benefit from an equally rising price due to the share of power that has been generated from fossil fuels. Second, the producers without regulated emissions may not have the market power to achieve higher selling prices with no additional regulatory (opportunity) costs. In order to test for these two competing hypotheses, regression Equation (1) is estimated for a control group. *Datastream* yields 12 entities making up all listed EU-based electricity firms that exclusively generate power with no regulated emissions.[41]

Table 19 reveals that the market β of the control sample yields a comparably higher value with a constant explanatory power. Neither the spot nor the futures

---

[41] It has to be noted that particularly the operators of solar and wind farms exhibit some differences to conventional electricity generators. Most of them are small with respect to number of employees and total assets pointing to lower capital intensity. Further, they are comparably young firms that are not listed in the main stock market segment at a small market capitalisation.

contract for pollution rights exhibit significance at conventional levels. This finding suggests that with respect to the electricity industry, investors in entities outside the regulation do not consider price movements on the allowance market. Only firms that are considerably engaged in emitting power generation are economically affected.

Table 19: Pricing Effects for Control Group

This table reports regression results of firm on market portfolio returns. Equation (1) is estimated as linear regression with White (1980) standard errors:

$$R_{it}^{ld} = \alpha + \beta_1 R_{mt}^{ld} + \beta_2 \, {}^{X}R_{et}^{ld} + \varepsilon \quad (1)$$

where: R_{it}^{ld} is the logged daily return on the common stock of firm i in period t; R_{mt}^{ld} is the logged daily return on the market portfolio; superscript $X = S$ denotes spot contracts, and $X = F$ denotes one year-ahead futures contracts; ${}^{S}R_{et}^{ld}$ is the logged daily return on emission allowances spot contracts; ${}^{F}R_{et}^{ld}$ is the logged daily return on emission allowances futures contracts mature in 2008-12. Two-tailed t-statistics are given in parentheses below a coefficient; R^2 is the adjusted coefficient of determination; F is the overall goodness of fit test. *, ** and *** denote levels of significance of 0.10, 0.05 and 0.01, respectively.

Independent Variables		Spot	Futures
Constant	α	0.000	0.000
		(1.356)	(1.235)
R_{mt}^{ld}	β_1	0.832***	0.830***
		(15.111)	(15.040)
${}^{X}R_{et}^{ld}$	β_2	0.004	0.018
		(1.510)	(1.514)
Number of Observations		8,964	8,964
R^2		0.0402	0.0402
F		114.73***	116.21***

4.5.2.2 Commodity Price Impact

In line with the idea of a multifactor market model, it is suggesting to add to the regression equation further variables which capture other potentially influencing factors. Since conventional producers depend on only a small number of input goods, returns on fuels will also affect future earnings. That is why a further examination should consider the impact of price changes in crude oil and natural gas. Especially the former factor has been tested in similar settings for different sectors and yields a strong relation with the stock market (Faff and Brailsford, 1999: 79-80; Sadorsky, 2001: 24; El-Sharif et al., 2005: 825). In this case, the intuition is simple: if fuel serves as an input factor then higher unit costs and thus lower earnings are anticipated, and vice versa. Both, oil and gas also drive emission allowance quotations: if, for example, the price of coal rises compared to gas, the latter will be preferred by power generators, resulting in lower emissions (Paolella and Taschini, 2008: 2023). Exogenous factors uniquely influencing emission rights may be due to modifications of the legal framework as well as weather changes (Alberola and Chevallier, 2007: 12-13; Benz and Trück, 2009: 6).

In the following analysis only crude oil and natural gas will be included as the most relevant factors likely to impact both stock and emission rights prices; the legal change is already incorporated in the futures. As oil and gas are invoiced in USD, the exchange rate to the local currency may also affect the power sector and becomes a further relevant variable.[42] An introduction of only two commodities is probably incapable of capturing all relevant economic input price effects. Hence, firm fixed effects are introduced that go beyond the impact of the other explanatory variables (de Jong et al., 2006: 23) and estimate Equation (4):[43]

$$R_{it}^{ld} = \alpha + \beta_1 R_{mt}^{ld} + \beta_2\, {}^{X}R_{et}^{ld} + \beta_3\, {}^{X}R_{ot}^{ld} + \beta_4\, {}^{X}R_{gt}^{ld} + \beta_5\, {}^{X}R_{ft}^{ld} + \varepsilon \tag{4}$$

where:

${}^{X}R_{ot}^{ld}$ = logged daily return on crude oil in period t;

${}^{X}R_{gt}^{ld}$ = logged daily return on Zeebrugge natural gas in period t;

${}^{X}R_{ft}^{ld}$ = logged daily return on local currency towards USD in period t;

X = $X = S$ denotes spot contracts, and $X = F$ denotes one-year-ahead futures/forward contracts.

---

42 The data for crude oil, natural gas and exchange rates are based on following contracts: the *London Brent Crude Index* spot in USD/barrel, the *International Commodity Exchange's ICE Brent Crude Oil* continuous future in USD/barrel, the average *Zeebrugge Natural Gas Day-Ahead* price in USD/Mmbtu, the average *Zeebrugge Natural Gas from October to September* as traded as futures in USD/MMBtu, the exchange rate of EUR/USD, GBP/USD or CZK/USD in spot and forward contracts. Quotations for oil and currencies were obtained from *Datastream*, those for gas were provided by *Platts*.

43 For the sake of parsimony, the coefficients for fixed effects are unreported in the equation and result tables.

Table 20: Pricing Effects of Commodities

This table reports regression results of firm on market portfolio returns. Equation (4) is estimated as linear regression including firm fixed effects with White (1980) standard errors:

$$R_{it}^{ld} = \alpha + \beta_1 R_{mt}^{ld} + \beta_2 \, {}^{X}R_{et}^{ld} + \beta_3 \, {}^{X}R_{ot}^{ld} + \beta_4 \, {}^{X}R_{gt}^{ld} + \beta_5 \, {}^{X}R_{ft}^{ld} + \varepsilon \qquad (4)$$

where: R_{it}^{ld} is the logged daily return on the common stock of firm i in period t; R_{mt}^{ld} is the logged daily return on the market portfolio; superscript $X = S$ denotes spot contracts, and $X = F$ denotes one year-ahead futures/forward contracts; ${}^{S}R_{et}^{ld}$ is the logged daily return on emission allowances spot contracts; ${}^{F}R_{et}^{ld}$ is the logged daily return on emission allowances futures contracts mature in 2008-12; ${}^{S}R_{ot}^{ld}$ is the logged daily return on the *London Brent Crude Index*; ${}^{F}R_{ot}^{ld}$ is the *ICE Brent Crude Oil* continuous year-ahead future; ${}^{S}R_{gt}^{ld}$ is the logged daily return on the average *Zeebrugge Natural Gas Day-Ahead* price; ${}^{F}R_{gt}^{ld}$ is the logged daily return on the average *Zeebrugge Natural Gas from October to September* futures; ${}^{S}R_{ft}^{ld}$ is logged daily change in the exchange rate for EUR/USD, GBP/USD or CZK/USD, respectively; and ${}^{F}R_{ft}^{ld}$ is the logged daily return on the exchange rate forward for EUR/USD, GBP/USD or CZK/USD, respectively. Two-tailed t-statistics are given in parentheses below a coefficient; R^2 is the adjusted coefficient of determination; F is the overall goodness of fit test. *, ** and *** denote levels of significance of 0.10, 0.05 and 0.01, respectively.

Independent Variables		Spot	Futures
Constant	α	0.000***	0.000***
		(14.135)	(13.645)
R_{mt}^{ld}	β_1	0.735***	0.712***
		(11.723)	(11.639)
${}^{X}R_{et}^{ld}$	β_2	0.004**	0.027***
		(2.440)	(4.915)
${}^{X}R_{ot}^{ld}$	β_3	0.039***	0.015**
		(2.870)	(2.062)
${}^{X}R_{gt}^{ld}$	β_4	0.001	0.001
		(0.624)	(0.077)
${}^{X}R_{ft}^{ld}$	β_5	0.240***	0.188***
		(8.012)	(7.249)
Number of Observations		19,362	19,362
R^2		0.1145	0.1142
F		30.25***	34.79***

Table 20 illustrates that the slope of the market portfolio lies in the same range for the spot and for the year-ahead estimation, but some 10% higher than in the reference model. Emission rights coefficients remained nearly unchanged. Both Brent contracts have a positive β for each specification. Surprisingly, the futures for pollution rights seem to have a larger impact on stock market returns since their economic effects is valued nearly twice as high. Natural gas contracts are not statistically different from zero. It can be supposed that the information incorporated in oil prices lead gas prices, and hence an inclusion of the latter does not exert an econometric effect. Further, the exchange rate employed here describes the relation between the USD and the local currency; that is, a positive return means a devaluation of the EUR, GBP or CZK, respectively. This positive relation strongly outnumbers the other additional parameters.

4.5.2.3 Implications of Power Generation

Apart from analysing issues relating to commodity price changes it is necessary to check for the robustness of the basic model with respect to important implications of power generation, especially production growth and the firm-specific fuel mix. Firstly, it is important to control for the growth in production, since an increased production ceteris paribus leads to higher regulated emissions. The cap installed by the regulator inevitably leads to a higher regulatory burden, either due to additionally needed abatement measures or due to the purchase of further certificates. In this respect, growth can bear a negative impact on stock valuation. The variable $DPRGRW_i$ captures the mean national annual growth in electricity production from 2004 through 2006 as published by *Eurelectric*. To be consistent with additional robustness checks, the values are coded as a dummy indicating 1 for growth above sample median and 0 otherwise.

Secondly, the proportion of power generated in regulated facilities, $DEMITT_i$, is measured. The intuition is that the impact of emission trading varies with a firm's dependence on conventional power generation. This indicator was estimated using data from financial reports, and equals 1 for entities with a share in emitting generation above sample median and 0 otherwise. Following prior literature, the impact of energy generation implications is included by employing interaction coefficients (Nydahl, 1999: 253; de Jong et al., 2006: 22-23):

$$R_{it}^{ld} = \alpha_1 + \alpha_2 DPRGRW_i + \alpha_3 DEMITT_i + \beta R_{mt}^{ld} + (\gamma_1 + \gamma_2 DPRGRW_i + \gamma_3 DEMITT_i)^X R_{et}^{ld} + \varepsilon \quad (5)$$

where:

$DPRGRW_i$ = dummy variable equalling 1 if the mean annual growth in electricity production in the country of incorporation of firm *i* is above sample median, and 0 otherwise;

$DEMITT_i$ = dummy variable equalling 1 if the share of fossil power generation for firm i is above sample median, and 0 otherwise.

Table 21: Pricing Effects of Energy Generation Implications

This table reports regression results of firm on market portfolio returns. Equation (5) is estimated as linear regression with White (1980) standard errors:

$$R_{it}^{ld} = \alpha_1 + \alpha_2 DPRGRW_i + \alpha_3 DEMITT_i + \beta R_{mt}^{ld} + (\gamma_1 + \gamma_2 DPRGRW_i + \gamma_3 DEMITT_i)\ {}^{X}R_{et}^{ld} + \varepsilon \quad (5)$$

where: R_{it}^{ld} is the logged daily return on the common stock of firm i in period t; R_{mt}^{ld} is the logged daily return on the market portfolio; superscript $X = S$ denotes spot contracts, and $X = F$ denotes one year-ahead futures contracts; ${}^{S}R_{et}^{ld}$ is the logged daily return on emission allowances spot contracts; ${}^{F}R_{et}^{ld}$ is the logged daily return on emission allowances futures contracts mature in 2008-12; $DPRGRW_i$ is a dummy variable equalling 1 if the mean annual growth in electricity production in the country of incorporation of firm i is above sample median, and 0 otherwise; $DEMITT_i$ is a dummy variable equalling 1 if the share of fossil power generation for firm i is above sample median, and 0 otherwise. Two-tailed t-statistics are given in parentheses below a coefficient; R^2 is the adjusted coefficient of determination; F is the overall goodness of fit test. *, ** and *** denote levels of significance of 0.10, 0.05 and 0.01, respectively.

Independent Variables		Spot	Futures
Intercept	α_1	0.001*** (3.498)	0.001*** (3.189)
$DPRGRW_i$	α_2	0.000 (-1.299)	0.000 (-0.955)
$DEMITT_i$	α_3	-0.000** (-2.031)	-0.000** (-2.069)
R_{mt}^{ld}	β	0.681*** (35.863)	0.677*** (35.814)
${}^{X}R_{et}^{ld}$	γ_1	0.007*** (3.118)	0.057*** (7.363)
$DPRGRW_i \times {}^{X}R_{et}^{ld}$	γ_2	-0.007*** (-3.503)	-0.033*** (-4.295)
$DEMITT_i \times {}^{X}R_{et}^{ld}$	γ_3	0.001 (0.528)	-0.020*** (-2.636)
Number of Observations		19,362	19,362
R^2		0.1098	0.1127
F		223.60***	224.90***

The results for these tests are reported in Table 21. The slopes for the market portfolio and allowance returns are similar in comparison to sign, magnitude and significance of Equation (1). The first interaction coefficient, the increase in power production, is significantly negative in both specifications. This result supports the hypothesis of a negative valuation impact of this issue. Firms facing growth seem to be able to generate only a small positive increment on the ETS ($\gamma_1 + \gamma_2$) for spot and futures contracts. The second interaction coefficient capturing the incremental valuation effect to big polluters behaves differently. It is insignificant with respect to spot contracts and significantly negative with respect to futures contracts. Investors predict that fuel choice does not play an influential role during the first phase whereas entities that rely more heavily on fossil fuels are perceived to face an additional burden in the second period.

4.5.2.4 Market and Regulatory Conditions

A fourth robustness check aims at controlling for variables that are likely to drive the results with respect to market and regulatory conditions. Despite several cross-border mergers, the share of the three biggest producers on most national wholesale markets (i. e. the respective concentration ratios) dropped slightly by 1% to 3% between the end of 2004 and the end of 2006 (European Commission, 2005: 44; European Commission, 2008: 12-13). Since the national markets have remained isolated and play the most important role in political and economic evaluation (Convery et al., 2008: 18), no significant impact of changes in competition on the main findings is predicted. The variable $DMKTCHNG_i$ captures the relative difference in national concentration ratios as indicated by the EU benchmark reports.[44] In order to reduce noise, it is coded as a dummy indicating 1 for a high reduction of concentration and 0 otherwise.

As to regulatory aspects, the European Commission put forward several directives to curb concentration and to furnish further liberalisation in the energy markets. Traditional approaches like long-term contracts or monopolies of local suppliers on the retail markets are to be replaced by polypolistic arrangements. To foster competition in retail markets, measures like price controls are regarded

---

44 Measuring market power by means of the three-firm concentration ratio probably does not capture the underlying economic effects inherent in energy markets fully. Residual Supply Indices (RSI) are considered a more reliable proxy yielding the percentage of electricity supplied to a grid not including firm *i*; hence a supplier's RSI indicates whether it is pivotal to the market. For a recent application of RSI to six European electricity markets see London Economics (2007). However this measure cannot be usefully applied in this study for data availability reasons. An estimation of these indices requires the knowledge of demand and capacity data at a high (usually hourly) periodicity that cannot be derived from publicly available information (Rahimi and Sheffrin, 2003: 490). Investors are unlikely to possess this kind of data.

as distorting and to be removed from the regulatory agenda (ERGEG, 2007: 6). Hence, a test for market competitiveness is incorporated by estimating the variable $DTARIFF_i$: it takes the value 1 if a regulatory authority is qualified to determine end user tariffs and 0 otherwise (ERGEG, 2007: 10).

A third variable is estimated capturing the endowment status of the sample firms. It is in the nature of a trading system that some firms are sellers and others buyers. If there are agents that pollute less than their endowment of certificates they will sell the exceeding rights and increase their financial wealth. Clearly, the distributional effects between the regulated installations depend on the level of initial allocation. The allocation status allows for controlling whether over-endowed firms are perceived to generate significant profits from selling allowances. For this reason the additional dummy variable $DENDOW_{it}$ is estimated, equalling = 1 for entities with an allocation exceeding the respective annual amount of verified emissions, and = 0 otherwise.

Regression equation (6) follows the previous specification with interaction terms capturing the market and regulatory impact on market valuation:

$$\begin{aligned} R_{it}^{ld} &= \alpha_1 + \alpha_2 DMKTCHG_i + \alpha_3 DTARIFF_i + \alpha_4 DENDOW_{it} + \beta R_{mt}^{ld} \\ &+ (\delta_1 + \delta_2 DMKTCHG_i + \delta_3 DTARIFF_i + \delta_4 DENDOW_{it})^X R_{et}^{ld} + \varepsilon \end{aligned} \tag{6}$$

where:

$DMKTCHG_i$ = dummy variable equalling 1 if the concentration ratio in the country of incorporation of firm *i* fell more than the sample median, and 0 otherwise;

$DTARIFF_i$ = dummy variable equalling 1 if the independent regulatory authority in the country of incorporation of firm *i* is qualified to determine end user prices, and 0 otherwise;

$DENDOW_{it}$ = dummy variable equalling 1 if firm *i* was allocated more rights than it emitted during an annual cycle, and 0 otherwise.

Table 22: Pricing Effects of Market and Regulatory Conditions

This table reports regression results of firm on market portfolio returns. Equation (6) is estimated as linear regression with White (1980) standard errors:

$$\begin{aligned} R_{it}^{ld} &= \alpha_1 + \alpha_2 DMKTCHG_i + \alpha_3 DTARIFF_i + \alpha_4 DENDOW_{it} + \beta R_{it}^{ld} \\ &+ (\delta_1 + \delta_2 DMKTCHG_i + \delta_3 DTARIFF_i + \delta_4 DENDOW_{it})\, {}^{X}R_{et}^{ld} + \varepsilon \end{aligned} \quad (6)$$

where: R_{it}^{ld} is the logged daily return on the common stock of firm i in period t; R_{mt}^{ld} is the logged daily return on the market portfolio; superscript $X = S$ denotes spot contracts, and $X = F$ denotes one year-ahead futures contracts; ${}^{S}R_{et}^{ld}$ is the logged daily return on emission allowances spot contracts; ${}^{F}R_{et}^{ld}$ is the logged daily return on emission allowances futures contracts mature in 2008-12; $DMKTCHG_i$ is a dummy variable equalling 1 if the concentration ratio in the country of incorporation of firm i fell stronger than the sample median, and 0 otherwise; $DTARIFF_i$ is a dummy variable equalling 1 if the independent regulatory authority in the country of incorporation of firm i is qualified to determine end user prices, and 0 otherwise; $DENDOW_i$ is a dummy variable equalling 1 if firm i was allocated more rights than it emitted during an annual cycle, and 0 otherwise. Two-tailed t-statistics are given in parentheses below a coefficient; R^2 is the adjusted coefficient of determination; F is the overall goodness of fit test. *, ** and *** denote levels of significance of 0.10, 0.05 and 0.01, respectively.

Independent Variables		Spot	Futures
Constant	α_1	0.001*** (2.717)	0.000** (2.227)
$DMTKCHG_i$	α_2	0.000 (-0.883)	0.000 (-0.770)
$DTARIFF_i$	α_3	0.000 (-0.759)	0.000 (-0.397)
$DENDOW_i$	α_4	0.000 (0.580)	0.000 (0.481)
R_{mt}^{ld}	β	0.681*** (35.845)	0.677*** (35.839)
${}^{X}R_{et}^{ld}$	γ_1	0.010*** (4.790)	0.055*** (6.959)
$DMKTCHG_i \times {}^{X}R_{et}^{ld}$	γ_2	-0.002 (-0.823)	-0.006 (-0.684)
$DTARIFF_i \times {}^{X}R_{et}^{ld}$	γ_3	-0.010*** (-4.060)	-0.035*** (-3.559)
$DENDOW_t \times {}^{X}R_{et}^{ld}$	γ_4	0.009*** (2.965)	-0.006 (-0.622)
Number of Observations		19,362	19,362
R^2		0.1107	0.1122
F		171.30***	169.00***

Table 22 shows the results for the market and regulatory variables. Both observations regarding the coefficient capturing market developments are insignificant; this conforms to the hypothesis that the changes in concentration ratios are too small to impact the fundamental findings. The end-user price controls further exhibit consistently negative slopes. This interaction term indicates the investors believe that entities still subject to tariff regulation will experience a negative effect on future income. The net effect of basic and dummy coefficients shows that price controls are perceived to almost fully exclude passing on the allowance cost. These results suggest that the European power markets do not seem to be as competitive as they would need to be to completely forego price controls. An analysis of the third variable, the effect of the endowment status, yields a mixed picture. On the one hand, over-endowed firms are predicted to generate net profits when controlling for allowance spot contracts, which is consistent with the hypothesis that generously allocated emitters are able to generate an additional profit in selling supernumerous rights. On the other hand, no significant result can be found for futures contracts. At first sight, the asymmetric significance towards both contract types is unsettling; at further inspection, it is evident that the endowment status is only connected to the spot contracts qualifying for a settlement of the present period's obligation.

4.5.3 Concluding Remarks

It was shown that the EU ETS has an economic effect at firm level: the certificates are regarded by investors as a relevant factor in valuation. However, this effect does not occur in the direction that regulators intended; generally shares in this industry are positively correlated to rising prices for emission allowances. Whereas spot quotations largely offer no significant contribution after the massive drop in their prices, futures exhibit a stable explanatory power in the course of time. This behaviour pinpoints a temporary lack of economic significance of the trading scheme, with expected positive wealth effects again in the second trading period.

The main findings remain unchanged when tested with several input factors. It is shown that the shares in this sector react positively towards a rising crude oil price and exchange rate, whereas natural gas prices and stock returns are not correlated. Additionally, for firms with a high share of fossil fuels employed in electricity generation, one third of the positive effect of emission trading is wasted in the second trading period due to their production restrictions. Investors in this sub-sample anticipate an economically adverse component once the trading scheme is fully in force. Recent changes in market structure did not impact the market valuation, as opposed to end-user price controls.

However, these findings have to be interpreted with caution. Firstly, they apply only to the electricity generating industry and are not necessarily valid for all other sectors involved. Once other sectors are included into the EU ETS, the dominance of the energy sector in the trading system might decline. Secondly, the overall positive correlation between allowance and stock returns does not hold for utilities with a high proportion of fossil electricity generation. Their special position within the industry will play an important role during the political process of determining future allocation modes.

4.6 Interacting Effects of Energy and Allowance Markets

4.6.1 Results

The previous analysis of question Q1 showed that the emission trade induced distributional effects during its first phase. Allowing for determinants like end user price controls yielded first evidence that the role of the state should not be neglected when estimating the quantity and quality of rents shifted. The full extent of public influence on energy markets cannot be sufficiently proxied by this single item of regulatory competence only; a broader approach of market regimes seems likely to better capture the underlying mechanisms and determinants for rents arising from the ETS. As set out, the full liberalisation of the European energy markets gained its effect in 2003 – hence it might be inappropriate to label the regimes as market-, state- and community-based. Since even the most recent assessment reports on the market transitions attest a slow progress for many Member States (European Commission, 2008), it may be more accurate to think in energy market traditions. Table 23 describes the distribution of sample firms over the three available regimes. It is apparent that entities of market and communitarian traditions account for roughly one quarter each, with a dominance of statist solutions. This is due to the fact that prior to (legal) full liberalisation, most countries pursued a state-based energy market approach.

Table 23: Energy Market Regimes of Sample Firms

The classification of countries into one of the three traditions corresponds to the predominant energy market regime prior to full liberalisation in 2003. An overview of the steps in market regime transition is given in Table 8.

Energy Market Regime	Countries	Number of Sample Firms
Market Tradition	Finland, United Kingdom	6
Statist Tradition	Czech Republic, France, Greece, Italy, Portugal, Spain	13
Communitarian Tradition	Austria, Germany	7
Sum		26

Table 24: Pricing Effects of Emission Allowances in Different Energy Market Regimes

This table reports regression results of firm on market portfolio returns. Equation (1) is estimated as linear regression with White (1980) standard errors:

$$R_{it}^{ld} = \alpha + \beta_1 R_{mt}^{ld} + \beta_2 \, {}^{F}R_{et}^{ld} + \varepsilon \qquad (1)$$

where: R_{it}^{ld} is the logged daily return on the common stock of firm i; R_{mt}^{ld} is the logged daily return on the market portfolio; ${}^{F}R_{et}^{ld}$ is the logged daily return on emission allowances futures contracts mature in 2008-12. Two-tailed t-statistics are given in parentheses below a coefficient; R^2 is the adjusted coefficient of determination; F is the overall goodness of fit test. *, ** and *** denote levels of significance of 0.10, 0.05 and 0.01, respectively.

Energy Market Regime	Constant α	R_{mt}^{ld} β_1	${}^{F}R_{et}^{ld}$ β_2	N	R^2	F
Market Tradition	0.000	0.774***	0.056***	4,386	0.1274	236.20***
	(1.286)	(21.013)	(6.301)			
Statist Tradition	0.000*	0.756***	0.015***	9,663	0.1464	437.60***
	(1.908)	(29.546)	(3.355)			
Communitarian Tradition	0.001**	0.450***	0.031***	5,313	0.0529	74.37***
	(2.562)	(11.463)	(4.102)			

Table 24 shows the results of estimating Equation (1) for each of the regimes separately. In line with the reference model, the allowance coefficient is significantly positive in all cases. Further, the absolute differences of β_2 are statistically different: unreported estimates of an equivalent but interacted regression model yield significant increments between the groups at the 5% level. Consequently, market participants predict positive ETS-related rents in favour of the producers, but quantitatively different with respect to the energy market regime. Power producers in the statist setting are marked with the lowest absolute value, followed by those in the communitarian and market settings each with higher parameters. As hypothesised, the state seems able to exert influence on the former group of firms and to restrict the regulatory rents. Interestingly, this control only mitigates the effect but does not lead to an internalisation of the burden (which would result in a negative coefficient). Further, to the author's knowledge, the results presented for corporatist markets exhibit the first capital market based test for this regime. Based on this evidence, the local power sectors over-compensate for a larger amount of the regulatory costs than in statist regimes. This is consistent with regional monopolies giving no incentive for firms to internalise the costs. Finally, in the market-based regime, the highest distributive effect can be observed – an indication for lacking maturity and competition in these arrangements.

Based on this simple test, it can be concluded that there is evidence for differential pricing effects, and hence differential rent distributions, connected to the energy market regime. Giving a suchlike answer to research question Q2 allows for one conclusion: the interaction of market-based arrangements in environmental and energy issues renders the ETS largely ineffective. Transforming both policies at (nearly) the same time gives rise to an adjustment effect at the expense of the electricity consuming industries and end users. Stated differently: economic incentives to avoid emissions can only be expected if state-based energy market regimes intervene or if veritably competitive power markets are created.

4.6.2 Additional Analyses

4.6.2.1 Implications of Energy Market Transformation

Two further analyses are conducted to understand the properties of energy market traditions better. Since first transitions in the European market regimes were initiated as early as 1989, these investigations cannot directly test for the implications of emission trading that started only 16 years later. Consequently, it is more to the point to ask into the changes of aggregate pricing effects connected to this process without explicitly controlling for an ETS impact. An extant prior literature focussed on gauging the capital market effect of regulated versus competitive market settings and employed a simple single factor market model of the following form (see only Peltzman, 1976; Clarke, 1980; Norton, 1985):

$$R_{it}^{ld} = \alpha + \beta R_{mt}^{ld} + \varepsilon. \tag{7}$$

Following its underlying hypothesis, a high degree of regulation "buffers" adverse effects on the income of regulated firms (Peltzman, 1976: 230). Hence, it is predicted that the stricter the regulation, the lower the coefficient of the market portfolio. It should be noted this model attracted criticism due to its simplicity, as e. g. Chen and Merville (1986), and Ahn and Thompson (1989) found counterintuitive results. Being aware of these issues, this analysis is only regarded as a limited test for the transitions in energy market regimes, and is to either complement or question the findings of the more robust multifactor model on the ETS.

As before, sample S1 comprising 26 power firms is analysed; the observation period starts on 1990-01-01 and ends on 2008-03-20. To distinguish between the early and full liberalisation levels, two sub-periods are defined: the former time frame stretches from 1990-01-01 until 2005-04-25 the inauguration of the EU ETS, and the second one covers the first trading phase capturing the regulatory change for energy markets. By intention, the periods were not chosen to be cut on 2003-01-01 the start of the European power market, to be consistent with the base models discussed in this study.

Table 25 summarises the pricing effects the early liberalisation period (labelled "1990-2005") as well of the full liberalisation periods ("2005-2008"), each depicted for the full sample and for the sub-samples attributable to one of the three market traditions. Panel A illustrates that the aggregate market risk for power generators increased slightly in the course of time. This finding is consistent with a decreasing regulation of this industry since investors predict firm-level earnings to be more volatile, and hence to be less easily forecast. A closer look to Panel B illustrates that this effect is largely driven by changes in market-based regimes, almost doubling the absolute coefficient during transition: the level of regulation seems to have declined considerably. It is counter-intuitive to find a quite high initial value of systematic risk for statist energy markets, as outlined in Panel C. This result may be driven by differences in liberalisation speeds among the Member States, as further analysed in the subsequent section.

Table 25: Pricing Effects of Market Regimes

This table reports regression results of firm on market portfolio returns. Equation (7) is estimated as linear regression with White (1980) standard errors:

$$R_{it}^{ld} = \alpha + \beta R_{mt}^{ld} + \varepsilon \qquad (7)$$

where: R_{it}^{ld} is the logged daily return on the common stock of firm i; R_{mt}^{ld} is the logged daily return on the market portfolio. Two-tailed t-statistics are given in parentheses below a coefficient; R^2 is the adjusted coefficient of determination; F is the overall goodness of fit test. *, ** and *** denote levels of significance of 0.10, 0.05 and 0.01, respectively. The observation sub-periods run from 1990-01-01 to 2005-04-24 and from 2005-04-25 to 2008-03-20, respectively.

Energy Market Regime	Constant α	R_{mt}^{ld} β	N	R^2	F
Panel A: Full Sample					
1990-2005	0.000***	0.516***	60,848	0.0506	2,162.00***
	(2.808)	(46.503)			
2005-2008	0.000***	0.680***	19,362	0.1083	1,277.00***
	(3.358)	(35.734)			
Panel B: Market Tradition					
1990-2005	0.000*	0.391***	11,158	0.0288	218.20***
	(1.797)	(14.772)			
2005-2008	0.000	0.780***	4,386	0.1183	436.00***
	(1.345)	(20.882)			
Panel C: Statist Tradition					
1990-2005	0.000	0.617***	30,232	0.0641	1,498.00***
	(1.114)	(38.699)			
2005-2008	0.000*	0.758***	9,663	0.1455	868.40***
	(1.933)	(29.469)			
Panel D: Communitarian Tradition					
1990-2005	0.000**	0.426***	19,458	0.0438	530.40***
	(2.444)	(23.031)			
2005-2008	0.001***	0.453***	5,313	0.0494	132.00***
	(2.603)	(11.487)			

Still, comparing the β with the one for the market-based regime shows no noteworthy difference in the full liberalisation period. This can be seen as evidence for a less strict regulatory climate also in the case of the classical statist regime. Finally, Panel D shows a low and nearly unchanged systematic risk for corporatist power firms. Even though this regulatory approach based on (centralised) negotiations has no tradition for public state interventions, it seems to ensure an effective buffering towards adverse shocks. Presenting results for statist and especially for corporatist market traditions is novel as it complements the prior findings largely focussing on differences in regulatory regimes in the US. Further research on the various market traditions, also controlling for adjustment effects in phases of transition, will be necessary, e. g. to provide for a more elaborated proxy for market regimes.

Comparing the aggregate effects of the available regulatory regimes with the ability to over-compensate for emission right prices results in a more differentiated picture. Market-based regimes showed signs of little regulation and at the same time a large potential for an ETS cost pass-through. For the sake of free markets, the regulatory authorities seem not to effectively intervene here. State-based arrangements have witnessed a vast catch-up in terms of deregulation, but selective instruments rested by the state seem to prevent excessive rents on allowances. Finally, the communitarian energy markets are (to date) immune to significant steps in liberalisation which can be equally found on the aggregate level of regulation as well concerning distributional emission right effects. This finding stands in contrast to previous literature arguing that state-based regimes bound to political processes react slowest (cf. Pierson, 2000: 490-491).

4.6.2.2 Differentiating Early and Late Liberalisation Adopters

As set out above, energy market transformation did not take place at an equal speed, especially considering the large statist tradition cluster. Some Member States still maintain a high degree of intervention, whereas others introduced liberalised elements into the domestic markets from the beginning. Consequently, it may be appropriate to further distinguish the speed of liberalisation when examining the aggregate level of transition. As a further analysis, the state-based sub-sample is further differentiated into early (producers in the Czech Republic, Italy, Portugal and Spain) and late adopters (firms from France and Greece).[45]

The regression estimates tabulated in Table 26 confirm the prior finding of an increasing systematic risk from the first to the fully deregulated observation period. However, both sub-samples differ: Firms in early liberalised state-based

[45] This clustering is in line with Table 8, showing groups with "decreasing state influence" and "strong state influence".

markets exhibit an already low regulatory degree (high market risk) in the first phase, as illustrated in Panel A; the increment to the subsequent period is small. In Member States transposing the deregulation directives with a time lag, the differential effect is more prominent (depicted in Panel B). Taking capital market risk as indicator for the regulatory energy market environment furnishes evidence for a steep decline in state interventions. This result further refines the first evidence found for statist energy markets. Major reforms took place in those regimes with the strongest state influence, whereas deregulation already had attained a considerable degree for early adopters before the full liberalisation came into effect.

Table 26: Pricing Effects of Differential Market Regime Transformation

This table reports regression results of firm on market portfolio returns. Equation (7) is estimated as linear regression with White (1980) standard errors:

$$R_{it}^{ld} = \alpha + \beta R_{it}^{ld} + \varepsilon \qquad (7)$$

where: R_{it}^{ld} is the logged daily return on the common stock of firm i; R_{it}^{ld} is the logged daily return on the market portfolio. Two-tailed t-statistics are given in parentheses below a coefficient; R^2 is the adjusted coefficient of determination; F is the overall goodness of fit test. *, ** and *** denote levels of significance of 0.10, 0.05 and 0.01, respectively. The observation sub-periods run from 1990-01-01 to 2005-04-24 and from 2005-04-25 to 2008-03-20, respectively.

Energy Market Regime	Constant α	R_{mt} β	N	R^2	F
Panel A: Statist Tradition, Early Liberalisation Adopters					
1990-2005	0.000	0.625***	29,354	0.0647	1,465.00***
	(0.988)	(38.275)			
2005-2008	0.000*	0.737***	7,590	0.1533	702.60***
	(1.897)	(26.506)			
Panel B: Statist Tradition, Late Liberalisation Adopters					
1990-2005	0.001	0.359***	878	0.0477	43.96***
	(1.194)	(6.630)			
2005-2008	0.000	0.829***	2,073	0.1285	178.00***
	(0.647)	(13.343)			

4.6.3 Concluding Remarks

The previous normative policy analysis argued that the concurrent transformation of environmental and energy market regimes in the EU was inspired by British market based approaches to regulation. Consequently, significant implications for the industry subject to it can be expected that potentially give rise to adjustment processes with unfavourable aggregate welfare effects. The empirical findings presented in this section allow for two substantial conclusions: Firstly, the transformation process affected the firms differently. Those entities initially embedded into strong state-based regimes are now in fact faced a less regulated and more market-oriented environment. Severe processes of adaption have changed their exposure towards the aggregate economic development, with still largely pending effects on market structure. At the same time, evidence for a stronger stickiness with regard to effective deregulation was found in communitarian regimes, both on firm and market levels. Secondly, the energy market traditions induce different reactions to rents arising from the EU ETS. Market-based approaches did not avoid the existence of regulatory rents, but instead fostered them. More effective means of intervention seem still to be in place in state-based regimes where only small distributional effects can be found.

The transition of energy market regime marks the state's retirement from its operational responsibilities. By means of liberalised markets, demand and supply are envisaged to result in more efficient prices than publicly determined or negotiated ones. However, the underlying industry structure prevents the creation of genuinely competitive markets at the national level as the present market-based regimes showed to account for the largest ETS-related rents. Taking this implication into account, a minimum of regulatory responsibility resting with public authorities cannot be given up if the markets for power and for allowances are to increase overall efficiency. This is due to the fact that mitigating an allowance cost pass-through is more effective in state- than in market-based regimes. A short- and mid-term implication for policy makers is that increasing effectiveness of the trading scheme is first of all bound to energy, and not allowance market reforms.

4.7 The Relative Usefulness of Accounting Treatments for Emission Rights[46]

4.7.1 Results

To deal with research question Q3 – if there is an accounting treatment giving rise to financial reporting items with a pricing according to their economic function –, in this section value relevance regression models are estimated for each of the available accounting treatments. ETS-related balance sheet items arising from either of the reporting alternatives are tested by means of a price model, properties of the emission earnings impact are estimated in a return regression. Since the disclosure of quantitative accounting information regarding the trading scheme can only be found in less than one fifth of the selected regulated firms, it was necessary to model how the different reporting procedures process the qualitative information. Using the thinly available empirical data directly supports the conclusions drawn based on the modelled inference, but a direct (non-modelled) approach is not feasible for the basic regressions due to the small number of observations.

To begin with, the balance sheet items are analysed by means of price regressions as outlined in Equation (2). Results are reported in Table 27. The basal coefficients on earnings ($EARN_{it}^{q}$) and book value of equity ($^{adj}BVE_{it}^{q}$) are significantly different from zero and comparable over all models.[47] Additionally, the coefficients on the emission liabilities are negative and significantly different from zero in all models where – following previous notation – the IFRIC 3 cost method is labelled as COST, the IFRIC 3 revaluation method as REV, the proposed full fair value amendment as FFV and the net approach of USofA 101 as NET. The allowance assets in the COST, REV and FFV treatments exhibit significantly positive signs. In line with the underlying model assumption allowances are immediately expensed and do not give rise to a testable asset for the NET alternative. All items in the price regressions thus behave in a plausible fashion, and all accounting treatments are able to reflect the underlying econom-

---

46 This section is based on parts of Veith et al. (2010).

47 Some coefficients on emission items in this and in the following tests have high absolute values. This has economic and econometric reasons: on the one hand, even small accounting components as resulting from applying the NET approach are likely to have a considerably high valuation effect. On the other hand, the modelled data can serve only as a proxy for empirical data (for similar issues see Healy et al., 2002: 697).

ics. Since the NET approach only reports a liability if a firm is short in allowances, and hence a part of the explanatory variable is equal to zero, the statistical estimation can be potentially biased. As proposed by Austin and Hoch (2004: 415), this effect is controlled for by repeating the analysis restricted to non-zero values, and find stable results.

Table 27: Value Relevance of Emission Rights Balance Sheet Items

This table reports regression results of stock price on reported financial data and modelled emission rights balance sheet items. Equation (2) is estimated as linear regression including quarter and industry fixed effects with White (1980) standard errors:

$$P^q_{it} = \varphi + \omega_1 EARN^q_{it} + \omega_2\, {}^{adj}BVE^q_{it} + \lambda ETS\_BSI^q_{it} + \varepsilon \qquad (2)$$

where: P^q_{it} is the price of the common stock ending one month after fiscal quarter end of firm i in period t; $EARN^q_{it}$ is quarterly earnings before extraordinary items; ${}^{adj}BVE^q_{it}$ is the book value of equity, adjusted for the modelled ETS balance sheet item; $ETS\_BSI^q_{it}$ is the modelled ETS-specific balance sheet item (either allowances asset or emission liability) according to one of the four alternative treatments. All modelled variables are winsorised at the 1% and 99% percentiles, share price and all accounting variables are scaled by the market value at fiscal quarter beginning. Two-tailed t-statistics are given in parentheses below a coefficient; R^2 is the adjusted coefficient of determination; F is the overall goodness of fit test. *, ** and *** denote levels of significance of 0.10, 0.05 and 0.01, respectively. The accounting treatments are coded as follows: COST is IFRIC 3 cost method; REV is IFRIC 3 revaluation method; FFV is the proposed full fair value amendment; NET is USofA 101.

Independent Variables		Asset: COST Model (2a)	Asset: REV, FFV Model (2b)	Liability: COST, REV, FFV Model (2c)	Liability: NET Model (2d)
Constant	φ	0.980*** (51.193)	0.983*** (50.839)	0.989*** (52.351)	0.994*** (51.838)
$EARN^q_{it}$	ω_1	0.798*** (2.985)	0.782*** (3.084)	0.807*** (3.199)	0.898*** (3.084)
${}^{adj}BVE^q_{it}$	ω_2	0.042* (1.714)	0.045* (1.758)	0.042* (1.720)	0.073*** (3.050)
$ETS\_BSI^q_{it}$	λ	1.601*** (5.361)	0.885*** (4.382)	-1.515*** (-5.007)	-10.001*** (-2.741)
Number of Observations		598	598	598	598
R^2		0.1445	0.1309	0.1490	0.1100
F		7.96***	7.41***	8.71***	6.21***

Table 28: Value Relevance of the Emission Rights Earnings Impact

This table reports regression estimates of stock returns on reported financial data and the modelled net emission rights earnings impact. Equation (3) is estimated as linear regression including quarter and industry fixed effects with White (1980) standard errors:

$$R_{it}^{\delta q} = \theta + \upsilon_1 {}^{adj}EARN_{it}^{q} + \upsilon_2 EARNCH_{it}^{q} + \kappa ETS\_EARN_{it}^{q} + \varepsilon \qquad (3)$$

where: $R_{it}^{\delta q}$ is the discrete buy-and-hold common stock return for a three-month period ending one month after the end of fiscal quarter t of firm i; ${}^{adj}EARN_{it}^{q}$ is the quarterly earnings before extraordinary items adjusted for the modelled ETS earnings impact; $EARNCH_{it}^{q}$ is the change in quarterly earnings before extraordinary items with respect to the same quarter of a year earlier; $ETS\_EARN_{it}^{q}$ is the modelled quarterly revenue or expense on emission trading. All modelled variables are winsorised at the 1% and 99% percentiles, all accounting variables are scaled by the market value at fiscal quarter beginning. Two-tailed t-statistics are given in parentheses below a coefficient; R^2 is the adjusted coefficient of determination; F is the overall goodness of fit test. *, ** and *** denote levels of significance of 0.10, 0.05 and 0.01, respectively. The accounting treatments are coded as follows: COST is IFRIC 3 cost method; REV is IFRIC 3 revaluation method; FFV is the proposed full fair value amendment; NET is USofA 101.

Independent Variables		Earnings: COST Model (3a)	Earnings: REV Model (3b)	Earnings: FFV Model (3c)	Earnings: NET Model (3d)
Constant	θ	0.029**	0.034***	0.031***	0.056***
		(2.558)	(2.764)	(2.602)	(4.941)
${}^{adj}EARN_{it}^{q}$	υ_1	0.853***	0.936***	0.948***	0.939***
		(3.555)	(3.611)	(3.608)	(3.632)
$EARNCH_{it}^{q}$	υ_2	0.073	0.071	0.095	0.063
		(0.732)	(0.759)	(1.017)	(0.683)
$ETS\_EARN_{it}^{q}$	κ	-1.527**	-0.014	2.330***	16.926***
		(-2.426)	(-0.015)	(2.660)	(3.590)
Number of Observations		598	598	598	598
R^2		0.0862	0.0637	0.0676	0.0953
F		5.17***	3.48***	3.27***	4.83***

Estimates for the return regression as outlined in Equation (3) are presented in Table 28. Again, the basal coefficients on earnings (${}^{adj}EARN_{it}^{q}$) and earnings changes ($EARNCH_{it}^{q}$) have the expected signs and magnitudes for all accounting treatments. With respect to the pricing of the emission trading impact on earnings ($ETS\_EARN_{it}^{q}$), three findings are important. First, the coefficients on $ETS\_EARN_{it}^{q}$ are positive and significantly different from zero in the NET and the proposed FFV models. Both accounting methods therefore capture the market estimation of the underlying economics in a suitable fashion. Second, the coefficient on $ETS\_EARN_{it}^{q}$ under the COST method is negative and significantly different from zero. This method leads to an unexpected pricing of the ETS revenues (gains). Third, the coefficient on $ETS\_EARN_{it}^{q}$ of the REV treatment is statistically insignificant. The latter two findings indicate that the COST and REV models lead to ETS income statement components which are hard to interpret. The result can be explained by the asymmetric recognition of asset and liability changes in both treatments: whereas the liability is measured at current cost in both cases, an increase in the allowance asset value is only revealed after derecognition in the case of applying COST, and is directly accrued to the revaluation reserve in the case of adopting REV. Apparently, the FFV and NET models both resolve this problem. In the former, accounting mismatches are mitigated since both the liability and the allowance asset are measured at fair value after initial recognition, and all revaluation amounts are recognised in income. The latter approach, NET, usually recognises either an asset or a liability, with accounting mismatches being less likely to occur. Hence, it is also less probable that subsequent measurement of these positions distorts earnings figures. The main income statement effects under the NET model thus can be traced to initially recognizing liabilities and to expensing purchased allowances (when being "used").

The market assessment of ETS-related balance sheet items and earnings components of the four accounting treatments is summarised in Table 29. Summarising the prior results is also suitable to answer research question Q3: all liabilities are priced negatively and the allowance assets – where applicable – are priced positively. Unexpected pricing of ETS income statement components are found for the COST and REV methods. As discussed, this finding can be explained by both approaches leading to accounting mismatches that distort income figures.

It can be additionally analysed whether one treatment dominates others by scrutinising the ETS-related regression coefficients. Both, the FFV and NET approaches consistently exhibit all estimated items in accordance with the predicted properties. REV shows one insignificant, and COST one unexpectedly priced component. Both approaches are therefore dominated. With these results, it is not possible to determine one dominating approach. The NET treatment conveys consistent information but focuses on the impact on the income statement – assets beyond amortisation are only accounted for in special cases. Gross approaches yield finer information for balance sheet items but only the FFV approach allows for relevant earnings at the same time.

Table 29: Summary of market valuation of ETS items

This table reports the sign of regression coefficients for the balance sheet items as estimated in Equation (2) and presented in Table 27 as well as for the earnings impact as estimated in Equation (3) and presented in Table 28. A parameter with a level of significance beyond 0.10 is indicated by 0. The accounting treatments are coded as follows: COST is IFRIC 3 cost method; REV is IFRIC 3 revaluation method; FFV is the proposed full fair value amendment; NET is USofA 101.

Accounting Treatment	Asset	Liability	Earnings
Expectation	+	-	+
COST	+	-	-
REV	+	-	0
FFV	+	-	+
NET	N/A	-	+

4.7.2 Additional Analyses

4.7.2.1 Controlling for Risk Factors

Leverage, size and the market-to-book ratio have been found to influence firm-specific risk and thus firm value in prior literature (Fama and French, 1995: 134). Particularly leverage and the market-to-book ratio, a rough indicator for growth, can drive the economic consequences of emission trading at the firm level.[48] For instance, leveraged firms might be constrained in advancing abatement measures as their ability to raise new funds is limited. This could expose leveraged firms more strongly to the intended economic effects of the ETS. Hence, entities with high debt-to-asset ratios may be adversely evaluated by investors.

The impact of growth options on firm value is not straightforward. On the one hand, an increased production ceteris paribus leads to higher emissions. The cap installed by the regulator inevitably leads to higher expenses, either due to additionally needed abatement measures or due to the purchase of further certificates. In this respect, growth can have a negative impact on future income. On the other hand, the predicted growth in turnover does not necessarily lead to

[48] As differences in size of the sample firms are rather small, the tests for an influence of size remains unreported. The tabulated results were left materially unchanged.

lower absolute profits even if the firms need additional cash outlays due to environmental regulation as these outlays can be over-compensated. One common approach for measuring growth options is the market-to-book ratio to capture the value of investment opportunities – the higher the ratio, the steeper the expected growth (Smith and Watts, 1992: 267). However, the market-to-book ratio has been criticised for being a too imprecise measure since it is said to be influenced by other factors. An alternative is to evaluate research and development expenditures as a more reliable proxy (Ahmed, 1994: 383; Jones et al., 2000: 25). This latter approach does, however, not seem to be applicable in the case of the heavy industries like cement or metal processing where such intangible assets are of less importance. Thus, leverage LEV_{it}^{q} is introduced next to the market-to-book ratio MTB_{it}^{q} as control variables into the price and return regressions and formulate Equations (8) and (9) as follows:

$$P_{it}^{q} = \varphi + \omega_1 EARN_{it}^{q} + \omega_2 {}^{adj}BVE_{it}^{q} + \lambda ETS\_BSI_{it}^{q} + \xi_1 LEV_{it}^{q} + \xi_2 MTB_{it}^{q} + \varepsilon \qquad (8)$$

$$\begin{aligned} R_{it}^{\delta q} = \theta + \upsilon_1 {}^{adj}EARN_{it}^{q} + \upsilon_2 EARNCH_{it}^{q} + \kappa ETS\_EARN_{it}^{q} \\ + \xi_1 LEV_{it}^{q} + \xi_2 MTB_{it}^{q} + \varepsilon \end{aligned} \qquad (9)$$

where:

LEV_{it}^{q}	= total debt divided by total assets for firm *i* in quarter *t*;
MTB_{it}^{q}	= market value at the end of quarter *t* divided by the book value of equity for firm *i*.

Estimates for the price regression can be found in Table 30. Controlling for leverage and market-to-book ratio improves the explanatory power but does not alter the main findings with respect to the ETS balance sheet item coefficients. In accordance with previous studies, the market-to-book ratio has a positive and significant coefficient indicating that anticipated growth does not give rise to an adverse valuation (Collins and Kothari, 1989: 168; Biddle and Seow, 1991: 222; Freeman and Tse, 1992: 205; Smith and Watts, 1992: 270; Marquardt and Wiedman, 2004: 319). Table 31 presents the results of the four separate return regressions controlling for leverage and growth. Again, the coefficients for the ETS earnings impact remain qualitatively unaltered for all models. Estimates for the main effects of leverage and growth are in line with the previous literature (Biddle and Seow, 1991: 222; Ahmed, 1994: 394; Barth et al., 1999: 406).

Table 30: Value Relevance of Emission Rights Balance Sheet Items (Control for Risk Factors)

This table reports regression results of stock price on reported financial data and modelled emission rights balance sheet items controlled for risk factors. Equation (8) is estimated as linear regression including quarter and industry fixed effects with White (1980) standard errors:

$$P^q_{it} = \varphi + \omega_1 EARN^q_{it} + \omega_2 {}^{adj}BVE^q_{it} + \lambda ETS\_BSI^q_{it} + \xi_1 LEV^q_{it} + \xi_2 MTB^q_{it} + \varepsilon \quad (8)$$

where: P^q_{it} is the price of the common stock ending one month after fiscal quarter end of firm i in period t; $EARN^q_{it}$ is quarterly earnings before extraordinary items; ${}^{adj}BVE^q_{it}$ is the book value of equity, adjusted for the modelled ETS balance sheet item; $ETS\_BSI^q_{it}$ is the modelled ETS-specific balance sheet item (either allowances asset or emission liability) according to one of the four alternative treatments; LEV^q_{it} is total debt divided by total assets; MTB^q_{it} is market value divided by book value of equity. All modelled variables are winsorised at the 1% and 99% percentiles, share price and all accounting variables are scaled by the market value at fiscal quarter beginning. Two-tailed t-statistics are given in parentheses below a coefficient; R^2 is the adjusted coefficient of determination; F is the overall goodness of fit test. *, ** and *** denote levels of significance of 0.10, 0.05 and 0.01, respectively. The accounting treatments are coded as follows: COST is IFRIC 3 cost method; REV is IFRIC 3 revaluation method; FFV is the proposed full fair value amendment; NET is USofA 101.

Independent Variables		Asset: COST Model (8a)	Asset: REV, FFV Model (8b)	Liability: COST, REV, FFV Model (8c)	Liability: NET Model (8d)
Constant	φ	0.780*** (11.891)	0.774*** (11.702)	0.782*** (12.095)	0.728*** (11.057)
$EARN^q_{it}$	ω_1	0.824*** (2.900)	0.803*** (2.942)	0.827*** (3.056)	0.866*** (2.849)
${}^{adj}BVE^q_{it}$	ω_2	0.218*** (3.960)	0.223*** (3.966)	0.219*** (4.060)	0.267*** (4.889)
$ETS\_BSI^q_{it}$	λ	1.617*** (5.437)	0.993*** (5.021)	-1.590*** (-5.370)	-10.589*** (-3.264)
LEV^q_{it}	ξ_1	-0.091 (-1.540)	-0.088 (-1.459)	-0.089 (-1.518)	-0.028 (-0.431)
MTB^q_{it}	ξ_2	0.078*** (5.487)	0.080*** (5.561)	0.079*** (5.600)	0.084*** (5.682)
Number of Observations		598	598	598	598
R^2		0.2461	0.2368	0.2535	0.2250
F		9.99***	9.45***	10.98***	8.23***

Table 31: Value Relevance of the Emission Rights Earnings Impact (Control for Risk Factors)

This table reports regression estimates of stock returns on reported financial data and the modelled net emission rights earnings impact controlled for risk factors. Equation (9) is estimated as linear regression including quarter and industry fixed effects with White (1980) standard errors:

$$R_{it}^{\delta q} = \theta + \upsilon_1 {}^{adj}EARN_{it}^{q} + \upsilon_2 EARNCH_{it}^{q} + \kappa ETS\_EARN_{it}^{q} + \xi_1 LEV_{it}^{q} + \xi_2 MTB_{it}^{q} + \varepsilon \quad (9)$$

where: $R_{it}^{\delta q}$ is the discrete buy-and-hold common stock return for a three-month period ending one month after the end of fiscal quarter t of firm i; ${}^{adj}EARN_{it}^{q}$ is the quarterly earnings before extraordinary items adjusted for the modelled ETS earnings impact; $EARNCH_{it}^{q}$ is the change in quarterly earnings before extraordinary items with respect to the same quarter of a year earlier; $ETS\_EARN_{it}^{q}$ is the modelled quarterly revenue or expense on emission trading; LEV_{it}^{q} is total debt divided by total assets; MTB_{it}^{q} is market value divided by book value of equity. All modelled variables are winsorised at the 1% and 99% percentiles, all accounting variables are scaled by the market value at fiscal quarter beginning. Two-tailed t-statistics are given in parentheses below a coefficient; R^2 is the adjusted coefficient of determination; F is the overall goodness of fit test. *, ** and *** denote levels of significance of 0.10, 0.05 and 0.01, respectively. The accounting treatments are coded as follows: COST is IFRIC 3 cost method; REV is IFRIC 3 revaluation method; FFV is the proposed full fair value amendment; NET is USofA 101.

Independent Variables		Earnings: COST Model (9a)	Earnings: REV Model (9b)	Earnings: FFV Model (9c)	Earnings: NET Model (9d)
Constant	θ	0.055	0.058	0.058	0.059*
		(1.539)	(1.579)	(1.547)	(1.672)
${}^{adj}EARN_{it}^{q}$	υ_1	0.987***	1.071***	1.087***	1.066***
		(4.051)	(4.057)	(3.993)	(4.040)
$EARNCH_{it}^{q}$	υ_2	0.061	0.058	0.084	0.049
		(0.690)	(0.692)	(1.030)	(0.592)
$ETS\_EARN_{it}^{q}$	κ	-1.493**	-0.207	2.595***	16.982***
		(-2.359)	(-0.225)	(2.897)	(3.535)
LEV_{it}^{q}	ξ_1	-0.096*	-0.097*	-0.096*	-0.055
		(-1.789)	(-1.761)	(-1.734)	(-1.042)
MTB_{it}^{q}	ξ_2	0.017**	0.017**	0.016**	0.016**
		(2.486)	(2.492)	(2.338)	(2.378)
Number of Observations		598	598	598	598
R^2		0.0966	0.0741	0.0772	0.1025
F		4.91***	3.52***	3.21***	4.42***

4.7.2.2 Controlling for Emission Price Changes

The unexpected pricing of the ETS earnings impact as to the COST and REV models suggests that both treatments suffer from accounting mismatches: As can be obtained from Panel A of Table 32, negative earnings impacts can either be driven by higher emission prices ceteris paribus leading to an increased liability or by lower emission prices leading to a decreased asset. These effects are not balanced due to differing measurement techniques applied to either of the balance sheet items.

Investigating into the valuation effects of reporting alternatives for emission rights hence necessitates distinguishing between the cases of income in- and decrements given positive and negative allowance returns, respectively. Marquardt and Wiedman (2004: 306) control for the value relevance of accounting losses by means of a dummy variable that is interacted with earnings in an additional response coefficient. Following this approach, $DNEG^q_{it}$ and respective interaction terms capturing the impact of ETS expenses are introduced. Further, employing a multifactor market model, Oberndorfer (2009: 1120) showed that investors react asymmetrically to positive and negative allowance returns. In line with these findings and given the hypothesised accounting mismatches bound to movements on the emission market, a second dummy variable DR^q_{et} and respective interactions controlling for allowance price decreases are introduced. The base return regression as specified in Equation (3) is thus augmented as follows:

$$\begin{aligned} R^{\delta q}_{it} &= \theta_1 + \theta_2 DNEG^q_{it} + \theta_3 DR^q_{et} + \theta_4 DNEG^q_{it} \times DR^q_{et} \\ &+ \upsilon_1 {}^{adj}EARN^q_{it} + \upsilon_2 EARNCH^q_{it} + \zeta_1 R^{\delta q}_{et} + \zeta_2 DR^q_{et} \times R^{\delta q}_{et} \\ &+ \kappa_1 ETS\_EARN^q_{it} + \kappa_2 DNEG^q_{it} \times ETS\_EARN^q_{it} \\ &+ \kappa_3 DR^q_{et} \times ETS\_EARN^q_{it} + \kappa_4 DNEG^q_{it} \times DR^q_{et} \times ETS\_EARN^q_{it} + \varepsilon \end{aligned} \tag{10}$$

where:

$DNEG^q_{it}$ = dummy variable equalling 1 for a negative ETS earnings pact during fiscal quarter t (i. e. $ETS\_EARN^q_{it} < 0$), and 0 otherwise;

DR^q_{et} = dummy variable equalling 1 for negative returns on emission allowances during fiscal quarter t (i. e. $R^{\delta q}_{et} < 0$), and 0 otherwise;

$R_{et}^{\delta q}$ = discrete return on emission allowances during fiscal quarter *t*.

In model (10), ζ_1 captures the average pricing effect of positive discretionary allowance returns ($R_{et}^{\delta q}$) and ζ_2 the average incremental pricing effect of negative discretionary allowance returns. κ_1 then captures the pricing of ETS-related revenues and κ_2 reports the incremental effect of net emission expenses. Further, the parameters κ_3 and κ_4 capture the additional impact of negative certificate returns. Panel B of Table 32 informs how to interpret the interacted coefficients.

Table 32: Allowance Returns and Accounting Consequences

Panel A depicts how underlying economic events of the ETS are mapped to accounting income. Panel B illustrates the interpretation of summarised effects resulting from the interaction regression coefficients δ_i presented in Equation (10):

$$\begin{aligned} R_{it}^{\delta q} = {} & \theta_1 + \theta_2 DNEG_{it}^q + \theta_3 DR_{et}^q + \theta_4 DNEG_{it}^q \times DR_{et}^q + \upsilon_1{}^{adj}EARN_{it}^q + \upsilon_2 EARNCH_{it}^q \\ & + \zeta_1 R_{et}^{\delta q} + \zeta_2 DR_{et}^q \times R_{et}^{\delta q} + \kappa_1 ETS\_EARN_{it}^q + \kappa_2 DNEG_{it}^q \times ETS\_EARN_{it}^q \\ & + \kappa_3 DR_{et}^q \times ETS\_EARN_{it}^q + \kappa_4 DNEG_{it}^q \times DR_{et}^q \times ETS\_EARN_{it}^q + \varepsilon \end{aligned} \tag{10}$$

where: κ_1: basal coefficient capturing the ETS earnings impact; κ_2: incremental effect of a negative ETS earnings impact; κ_3: incremental effect of negative allowance returns; κ_4: incremental effect both a negative ETS earnings impact and negative allowance returns.

Panel A: Mapping of Allowance Returns to Accounting

		Accounting Event	
		Positive Effect	Negative Effect
Economic Event	Positive Allowance Returns	Increase in allowance asset (limited if measurement applies historical costs or revaluation method)	Increase in emission liability
Economic Event	Negative Allowance Returns	Decrease in emission liability	Decrease in allowance asset

Panel B: Interpretation of Interaction Regression Coefficients

		Accounting for Emission Rights	
		Base Case	Net Accounting Loss
Allowance Returns	Base Case	κ_1 Positive earnings impact Positive allowance returns	$\kappa_1 + \kappa_2$ Negative earnings impact Positive allowance returns
Allowance Returns	Negative Returns	$\kappa_1 + \kappa_3$ Positive earnings impact Negative allowance returns	$\kappa_1 + \kappa_2 + \kappa_3 + \kappa_4$ Negative earnings impact Negative allowance returns

Table 33: Value Relevance of the Emission Rights Earnings Impact (Interacted Return Models)

This table reports regression estimates of stock returns on reported financial data and the modelled net emission rights earnings impact controlled for emission rights accounting and returns asymmetries. Equation (10) is estimated as linear regression including quarter and industry fixed effects with White (1980) standard errors:

$$R_{it}^{\delta q} = \theta_1 + \theta_2 DNEG_{it}^q + \theta_3 DR_{et}^q + \theta_4 DNEG_{it}^q \times DR_{et}^q + \upsilon_1\,{}^{adj}EARN_{it}^q + \upsilon_2 EARNCH_{it}^q$$
$$+ \zeta_1 R_{et}^{\delta q} + \zeta_2 DR_{et}^q \times R_{et}^{\delta q} + \kappa_1 ETS\_EARN_{it}^q + \kappa_2 DNEG_{it}^q \times ETS\_EARN_{it}^q \qquad (10)$$
$$+ \kappa_3 DR_{et}^q \times ETS\_EARN_{it}^q + \kappa_4 DNEG_{it}^q \times DR_{et}^q \times ETS\_EARN_{it}^q + \varepsilon$$

where: $R_{it}^{\delta q}$ is the discrete buy-and-hold common stock return for a three-month period ending one month after the end of fiscal quarter t of firm i; $DNEG_{it}^q$ is a dummy variable equalling 1 indicating a emission expense; DR_{et}^q is a dummy variable equalling 1 for negative returns on emission allowances; ${}^{adj}EARN_{it}^q$ is the quarterly earnings before extraordinary items adjusted for the modelled ETS earnings impact; $EARNCH_{it}^q$ is the change in quarterly earnings before extraordinary items with respect to the same quarter of a year earlier; $R_{et}^{\delta q}$ is the discrete emission allowance return during the fiscal quarter; $ETS\_EARN_{it}^q$ is the modelled quarterly revenue or expense on emission trading. All modelled variables are winsorised at the 1% and 99% percentiles, all accounting variables are scaled by the market value at fiscal quarter beginning. Two-tailed t-statistics are given in parentheses below a coefficient; R^2 is the adjusted coefficient of determination; F is the overall goodness of fit test. *, ** and *** denote levels of significance of 0.10, 0.05 and 0.01, respectively. The accounting treatments are coded as follows: COST is IFRIC 3 cost method; REV is IFRIC 3 revaluation method; FFV is the proposed full fair value amendment; NET is USofA 101. Constant terms remain unreported for the sake of parsimony.

Independent Variables		Earnings: COST Model (10a)	Earnings: REV Model (10b)	Earnings: FFV Model (10c)	Earnings: NET Model (10d)
${}^{adj}EARN_{it}^q$	υ_1	0.672*** (3.037)	0.718*** (3.167)	0.782*** (3.196)	0.674*** (2.999)
$EARNCH_{it}^q$	υ_2	-0.001 (-0.005)	0.031 (0.310)	0.044 (0.433)	0.021 (0.196)
$R_{et}^{\delta q}$	ζ_1	0.000 (-1.572)	0.000 (-1.567)	-0.000** (-2.160)	-0.000** (-2.160)
$DR_{et}^q \times R_{et}^{\delta q}$	ζ_1	0.200*** (6.435)	0.189*** (5.260)	0.203*** (5.306)	0.167*** (5.450)
$ETS\_EARN_{it}^q$	κ_1	-8.452** (-2.171)	-10.678** (-2.316)	3.502** (2.170)	53.120*** (2.695)
$DNEG_{it}^q \times ETS\_EARN_{it}^q$	κ_2	6.641 (1.598)	8.659* (1.795)	-9.359*** (-3.343)	-46.895** (-2.208)
$DR_{et}^q \times ETS\_EARN_{it}^q$	κ_3	8.756** (2.160)	19.433*** (3.911)	3.954* (1.736)	-12.464 (-0.515)
$DNEG_{it}^q \times DR_{et}^q \times ETS\_EARN_{it}^q$	κ_4	4.983 (1.055)	-9.159* (-1.719)	5.148 (1.307)	17.774 (0.695)
Number of Observations		598	598	598	598
R^2		0.2158	0.2184	0.1617	0.1812
F		10.31***	11.69***	16.05***	6.88***

Table 34: Summary of Interactions for Emission Trade Related Earnings Components

This table illustrates the summarised effects of the interaction regression coefficients κ_i presented in Equation (10):

$$R_{it}^{\delta q} = \theta_1 + \theta_2 DNEG_{it}^q + \theta_3 DR_{et}^q + \theta_4 DNEG_{it}^q \times DR_{et}^q + \upsilon_1 {}^{adj}EARN_{it}^q + \upsilon_2 EARNCH_{it}^q + \zeta_1 R_{et}^{\delta q} + \zeta_2 DR_{et}^q \times R_{et}^{\delta q} + \kappa_1 ETS\_EARN_{it}^q + \kappa_2 DNEG_{it}^q \times ETS\_EARN_{it}^q + \kappa_3 DR_{et}^q \times ETS\_EARN_{it}^q + \kappa_4 DNEG_{it}^q \times DR_{et}^q \times ETS\_EARN_{it}^q + \varepsilon \quad (10)$$

where: κ_1: basal coefficient capturing the ETS earnings impact; κ_2: incremental effect of a negative ETS earnings impact; κ_3: incremental effect of negative allowance returns; κ_4: incremental effect both a negative ETS earnings impact and negative allowance returns. The accounting treatments are coded as follows: COST is IFRIC 3 cost method; REV is IFRIC 3 revaluation method; FFV is the proposed full fair value amendment; NET is USofA 101. The full results are presented in Table 33.

	Base Case	Net Accounting Loss
Panel A: Earnings COST (Model (10a))		
Base Case	$\kappa_1 =$ -8.5	$\kappa_1 + \kappa_2 =$ -8.5
Negative Allowance Returns	$\kappa_1 + \kappa_3 =$ +0.30	$\kappa_1 + \kappa_2 + \kappa_3 + \kappa_4 =$ +0.30
Panel B: Earnings REV (Model (10b))		
Base Case	$\kappa_1 =$ -10.7	$\kappa_1 + \kappa_2 =$ -2.0
Negative Allowance Returns	$\kappa_1 + \kappa_3 =$ +8.8	$\kappa_1 + \kappa_2 + \kappa_3 + \kappa_4 =$ +8.3
Panel C: Earnings FFV (Model (10c))		
Base Case	$\kappa_1 =$ +3.5	$\kappa_1 + \kappa_2 =$ -5.9
Negative Allowance Returns	$\kappa_1 + \kappa_3 =$ +7.5	$\kappa_1 + \kappa_2 + \kappa_3 + \kappa_4 =$ -1.9
Panel D: Earnings NET (Model (11d))		
Base Case	$\kappa_1 =$ +53.1	$\kappa_1 + \kappa_2 =$ -46.9
Negative Allowance Returns	$\kappa_1 + \kappa_3 =$ +53.1	$\kappa_1 + \kappa_2 + \kappa_3 + \kappa_4 =$ -46.9

Estimating model (10) first reveals that the allowance return coefficient ζ_j is overall positively valued by the market, which is in line with the findings presented in this study regarding the capital market pricing effects of the ETS. The basal emissions earnings impact coefficients κ_1 are largely consistent with the ones in the non-interacted Equation (3): the COST method remains with a negative, FFV and NET with a positive pricing. However, due to controlling for the sign of the earnings impact and of allowance returns, the parameter for REV becomes significantly negative. It now takes the same (negative) direction as the other IFRIC 3 approach, COST. Further, at least one of the respective interacted emission coefficients is statistically different from zero, revealing that controlling for non-linearity adds relevant and otherwise omitted information. Estimates are tabulated in Table 33.

Table 34 juxtaposes the pricing effects for emission net profits and losses in case of positive and negative allowance returns. Panel A depicts the case of the COST method revealing that decreases in allowance prices always lead to positive valuation effects, irrespective of the earnings impact. A negative effect holds for the opposite case. The findings suggest that the pricing of income components is rather driven by allowance price changes than by the respective revenues or expenses recorded in the income statement. This also extends to the REV treatment as can be seen in Panel B of Table 34 supporting the first evidence that both gross approaches suffer from accounting mismatches with distorted valuation effects. Panel C presents the results for the FFV proposal, and shows summarised slope coefficients with positive (negative) signs for net profits (losses) irrespective of allowance returns. Finally, the summarised coefficients for the NET treatment are sketched in Panel D. Again, the capital market valuation effect is only due to the sign of the ETS earnings impact, and equally consistent with the framework's assumptions towards accounting information properties.

4.7.2.3 Comparing Modelled with Empirical Data

Throughout the preceding analysis comparing the income statement and balance sheet effects of the different accounting treatments, modelled instead of real data were used. Modelling the items might introduce a bias to this analysis: the model, for instance, fails to generate an asset under the NET treatment, which is empirically observable when a firm has purchased allowances that have not yet been "used". Thus, price and return models with empirical data are estimated, where empirical data is obtained from the small subsample of firms disclosing the necessary data (results thus have to be interpreted cautiously), and the sign and significance of the coefficients is contrasted for real and modelled ETS-related components. The small sample size forces to perform an econometric modification to the regression, which does not alter the economic perspective. To arrive at an efficient estimation, the available observations of the two empirical subsamples containing IFRIC 3 (COST/REV) and USofA (NET) adopters are pooled and the additional IFRIC information is beckoned out. This is accomplished by interacting the respective ETS coefficients with a dummy variable.

In Equations (11) and (12), the coefficients $ETS\_BSI^q_{it}$ and $ETS\_EARN^q_{it}$ inform on the pricing of the respective NET items. Additionally, the dummy variable $DIFIRC^q_{it}$ is introduced taking the value one if a firm accounts for emission rights according to the COST/REV methods, and zero otherwise. The regressions take the following form:

$$P^q_{it} = \varphi_1 + \varphi_2 DIFIRC^q_{it} + \omega_1 EARN^q_{it} + \omega_2 {}^{adj}BVE^q_{it} + \lambda_1 ETS\_BSI^q_{it} + \lambda_2 DIFIRC^q_{it} \times ETS\_BSI^q_{it} + \varepsilon \tag{11}$$

$$R^{\delta q}_{it} = \theta_1 + \theta_2 DIFIRC^q_{it} + \upsilon_1 {}^{adj}EARN^q_{it} + \upsilon_2 EARNCH^q_{it} + \kappa_1 ETS\_EARN^q_{it} + \kappa_2 DIFIRC^q_{it} \times ETS\_EARN^q_{it} + \varepsilon \tag{12}$$

where:

$DIFIRC^q_{it}$ = dummy variable equalling 1 for IFRIC 3 adopters, and 0 otherwise.

The interacted model is interpreted as follows: coefficient λ_1 captures the pricing for the NET asset (liability) and the sum of $\lambda_1 + \lambda_2$ the one for the COST/REV asset (liability). Analogously, κ_1 informs on the increment to earnings following the NET method and the sum of $\kappa_1 + \kappa_2$ on the one for the COST/REV treatment in the returns regression. The results presented in Table 35 indicate that all regressions based on empirical data yield qualitatively the same findings as the modelled ones. Panel A shows the results for the assets from the price regression. Whereas the NET asset is not significantly different from zero, the summarised impact for the COST/REV asset is significantly positive, and is in line with both economic expectations and the prior modelling. Further, the liabilities as exhibited in Panel B show negative net effects for both empirical alternatives and are priced with the same sign as their modelled counterparts in Table 27. Finally, the valuation effects for the empirical earnings data presented in Panel C are consistent with modelling: the NET increment to

Table 35: Value Relevance of Empirical Emission Rights Financial Reporting Items

This table reports regression results of stock price or stock returns on reported financial data. Equations (11) and 12) are estimated as linear regressions including quarter and industry fixed effects with White (1980) standard errors:

$$P_{it}^{q} = \varphi_1 + \varphi_2 DIFIRC_{it}^{q} + \omega_1 EARN_{it}^{q} + \omega_2 {}^{adj}BVE_{it}^{q} + \lambda_1 ETS\_BSI_{it}^{q} + \lambda_2 DIFIRC_{it}^{q} \times ETS\_BSI_{it}^{q} + \varepsilon \quad (11)$$

$$R_{it}^{\delta q} = \theta_1 + \theta_2 DIFIRC_{it}^{q} + \upsilon_1 {}^{adj}EARN_{it}^{q} + \upsilon_2 EARNCH_{it}^{q} + \kappa_1 ETS\_EARN_{it}^{q} + \kappa_2 DIFIRC_{it}^{q} \times ETS\_EARN_{it}^{q} + \varepsilon \quad (12)$$

where: P_{it}^{q} is the price of the common stock ending one month after fiscal quarter end of firm i in period t; $R_{it}^{\delta q}$ is the discrete buy-and-hold common stock return for a three-month period ending one month after the end of fiscal quarter; $DIFIRC_{it}^{q}$ is a dummy variable, equal to 1 for IFRIC 3 adopters, and 0 otherwise; $EARN_{it}^{q}$ is the quarterly earnings before extraordinary items adjusted for the ETS earnings impact; ${}^{adj}BVE_{it}^{q}$ is the book value of equity, adjusted for the ETS balance sheet item; $EARNCH_{it}^{q}$ is the change in quarterly earnings before extraordinary items with respect to the same quarter of a year earlier; $ETS\_BSI_{it}^{q}$ is the empirical ETS-specific balance sheet item (either allowances asset or emission liability); $ETS\_EARN_{it}^{q}$ is the empirical quarterly revenue or expense on emission trading. Share price and all accounting variables are scaled by the market value at fiscal quarter beginning. Two-tailed t-statistics are given in parentheses below a coefficient; R^2 is the adjusted coefficient of determination; F is the overall goodness of fit test. *, ** and *** denote levels of significance of 0.10, 0.05 and 0.01, respectively.

	Constant	$DIFIRC_{it}^{q}$	$EARN_{it}^{q}$	BVE_{it}^{q}	$ETS\_BSI_{it}^{q}$	$DIFIRC_{it}^{q} \times ETS\_BSI_{it}^{q}$	N	R^2	F
Panel A: Value Relevance of Emission Rights Assets (Price Model)									
	φ_1	φ_2	ω_1	ω_2	λ_1	λ_2			
Asset: Model (11a)	1.095***	-0.164***	0.777*	0.062	-7.298	20.220**	120	0.1629	3.12***
	(17.461)	(-3.030)	(1.888)	(0.846)	(-1.030)	(2.489)			
Panel B: Value Relevance of Emission Liabilities (Price Model)									
Liability: Model (11b)	0.970***	-0.084	0.752*	0.049	-1.573*	-9.208***	141	0.1168	3.21***
	(14.189)	(-1.482)	(1.850)	(0.815)	(-1.965)	(-2.644)			
Panel C: Value Relevance of the Emission Rights Earnings Impact (Return Model)									
	Constant	$DIFIRC_{it}^{q}$	$EARN_{it}^{q}$	$EARNCH_{it}^{q}$	$ETS\_EARN_{it}^{q}$	$DIFIRC_{it}^{q} \times ETS\_EARN_{it}^{q}$	N	R^2	F
	θ_1	θ_2	υ_1	υ_2	κ_1	κ_2			
Earnings: Model (12)	0.041*	-0.010	0.365	0.639***	12.033*	-20.231*	118	0.7857	4.08***
	(1.818)	(-0.366)	(1.264)	(3.438)	(1.846)	(-1.659)			

income is priced positively, the summarised effect for the gross approach negatively, qualitatively corresponding to the COST model of Table 28. Taken together, the empirical evidence suggests that the applied model works well in producing results for the underlying accounting treatments, and that the conclusions are supported by the available real data.

4.7.3 Implications for Accounting Standard Setting

4.7.3.1 Final Evaluation of Effective and Proposed Accounting Treatments

The nature and complexity of an emission trading scheme pushes accounting to its limits. It induces substantial accounting problems for the regulated firms, which have not yet been satisfactorily resolved. To date, a large range of alternative accounting treatments exists. While prior literature has discussed technical properties of the different treatments, no evidence has been presented whether the different treatments give rise to economically meaningful balance sheet and income statement items and whether one of them seems superior to others. The results given here indicate that there are qualitative differences between treatments: the two accounting treatments outlined in IFRIC 3 (COST, REV) lead to income statement components which are not priced according to the expectations. A likely explanation is that both approaches distorting earnings figures by introducing accounting mismatches. Consistent pricing of earnings can be found for the full fair value approach proposed in the IAS 38 revision and the net approach used in USofA. Both produce earnings figures that perform equally well in explaining returns. However, taking the balance sheet issue into account, the discussed full fair value model (FFV) delivers finer information than the USofA (NET) model, and it is the only approach with complete and consistent asset, liability and income component pricing.

Findings are robust with respect to controlling for risk factors. Moreover, modelled data used throughout the analysis seem to proxy sufficiently well for the empirical data. Some limitations need to be acknowledged. It was necessary to rely on a relatively small sample as the EU ETS applies to a comparatively small number of listed firms and has only been in operation since 2005. The modelled data can possibly introduce noise to the analysis. Finally, the general shortcomings inherent with the value relevance research also apply here, especially with respect to econometric issues of price models. Additional analyses will have to show whether the inference presented here can be regarded as stable over the first emission trading period and when using empirical data.

4.7.3.2 Policy Implications

The previous analysis of modelled and empirical data showed that a full fair value amendment to an itemised accounting approach for emission rights is able to mitigate the distorting information effects in a capital market valuation setting. These findings should be considered in the ongoing deliberations at the IASB and the FASB concerning the future joint accounting standard Emission Trading Schemes. It is straightforward to acknowledge that basing the allowance asset as well as the emission liability on the same measurement technique (full fair value through profit or loss). This allows for the fair representation of the respective balance sheet items and promotes an ETS-related earnings impact netted by opposite in- or decrements in the asset and liability. At further inspection, reducing noise in income is also bound to aligning underlying quantities – the number of allowances held versus the level of emissions incurred. Keeping in mind these implications, a slightly refined way of treating emission rights should be considered that has (to the author's knowledge) yet to be discussed:

1. Allowances are measured at full fair value through profit or loss, as proposed by the IFRIC.
2. The liability corresponds to the projected emissions of the current trading cycle. That is, according to the best economic estimate, a liability is accrued as early as the allocation date. Changes in the liability arise on price changes (conceptually corresponding to the asset) or when the prior estimate is revised. For the sake of simplicity, suchlike estimations can be based on historical emissions for incumbent firms. In its early discussions, the IFRIC concluded that this approach violates IAS 37.14(a) and 17(a) that require a liability to arise from past events, and that there is no obligation as long as there have no emissions occurred (IFRIC 3 Basis for Conclusions, para. 22). Current discussion papers regard a voluntary or compulsory participation in a trading scheme as entering into an agreement giving rise to a grant that is due at surrender (IASB, 2009: para. 63).

Extending this intuition of going concern is consistent as accruing a liability based on estimated emissions can thus be deduced from the contract agreed on. If a firm plans to close its operations instead of operating under the regulation, no liability will be recognised.

3. A grant is only recognised for allowances allocated for consecutive trading cycles. Inline with IAS 41 and present deliberations, this liability is measured at full fair value.

This allows distinguishing over- (under-)endowed entities by means of day-one profits (losses) at the earliest possible date. With respect to the economic exposure towards a running trading cycle, this treatment is consistent with finer information than disclosing a shortfall only when it occurs as e. g. USofA does. Further net effects on interim earnings only arise to changes in the net value of the allowances held and of the estimated emission liability. Due to the full fair value measurement, rising allowance prices lead to losses (earnings) if there are less (more) allowances to cover predicted emissions, and vice versa.

It should be noted that even this itemised approach is also not able to fully map the economic mechanisms related to an emission trading scheme. This quantitative disclosure does not report on the distributive effects between the entity and its customers – as dealt with in answering research question Q1. Further, it remains unknown to investors in which way the investment activity or turnover are affected by the regulation. For this reason, a future accounting standard for emission rights also needs to include provisions for a qualitative disclosure: Firstly, it should be explained how the ETS affects the product policy in terms of selling volumes and prices, i. e. the net of passing on or bearing (opportunity) costs. Secondly, it should be illustrated to which degree investments into property, plant and equipment are triggered by the ETS.

As a final implication for the standard setter, it should be addressed why emission trading attracts a high attention in the energy economics, accounting, political, legal and engineering sciences literature, but its effect is only rarely separately described in financial reports. Based on the results presented in this study, this lack of disclosure cannot be justified by a lack of overall economic significance. That is why clear requirements should be laid out in which case the ETS-related quantitative and qualitative information needs to be disclosed in the notes. A determination of relevance can be bound to e. g. the ratio of emission liability vis-à-vis total assets or of emission earnings vis-à-vis total earnings.

5 Conclusions

The aim of this study was to analyse the institutional framework of and economic consequences arising from the EU Emission Trading Scheme. In doing so, a review on the transformation of environmental and energy policies in the European Union as well as a normative analysis of mapping emission tights to management and financial accounting were provided. Finally, three empirical analyses were performed furnishing evidence for distinct economic consequences arising from this regulation.

Basis of this study was a governance framework that develops an institutionalist concept of environmental and energy policies. Unlike prior research, this taxonomy asks for the actors providing welfare, and isolates the state, the community and the market as alternative solutions. Traditional environmental policy instruments rely on state interventions, prescribing technology or performance standards, subsidies or taxes. Even though all of these measures induce economic incentives – e. g. imposing a tax on certain goods, which leads to a differentiation in market prices – the (re)distribution of welfare rests within the competence of a public agency. A communitarian approach of environmental regulation consists of compensation or negotiated agreements between polluters and their victims. Completely differently works an ETS that develops its regulatory character by means of its endogenous pricing mechanism. The state remains responsible for setting up and enforcing the framework, but the desired incentivising economic effect eventuates beyond the influence of the regulator.

The overview given further sketched the implementation of this market-based instrument for the case of the EU. Due to the dispersed competences of the Member States and the Commission, rules for determining the emission budget to be allocated turned out to be too generous during the first three trading years (2005-2007). Several national governments achieved in granting more allowances than ex post physically emitted, especially to those sectors (potentially) exposed to carbon leakage, like metals, minerals and paper processing. Finally, this study discussed why a novel instead of a traditional policy instrument was employed. The EU emerged as actor on this policy field instead of the nation state since capping greenhouse gas emissions can only be effectively accomplished on a supranational level. With no own fiscal competences, the scope of the Union was ex ante restricted to non-financial measures. Additionally, the postulate of economic efficiency entered the political decision making process and paved the way for a trading scheme.

Similar developments can be observed with respect to the European energy market regimes. Applying once again the institutionalist taxonomy, this study differentiated state-, community- and market-based arrangements. If energy generation and transmission is carried out by SOEs, the state acts as provider of welfare. This constituted the predominant market regime in most European countries during the golden age of the nation state until the end of the 1980s. Another traditional regime relies on the corporatist oligopolisation of energy generation with limited public interference. In 1989 and 2003, the European energy markets were liberalised in two steps, promoting market-based solutions. This study illustrated that the transformation of policy instruments was justified by the competence of the EU to create a single market, and additionally by introducing economic efficiency and competition to these separated markets. The contemporaneous and analogous developments in environmental and energy policies raised two questions: first, does the introduction of a market-based emission regulation lead to economic rents?, and second, does the prevailing energy market regime determine the extent or direction of these rents?

This analysis further investigated into the economic consequences of an ETS on reporting instruments at the business level. If the environmental regulation is set out to gain an incentivising effect, this has to be reflected in management and financial accounting. Both information channels have similar functions, namely stewardship and decision making, but their perspectives and implementations differ. Management accounting is mainly intended for the internal evaluation of divisions or of the management board: it is not strictly codified and the accountant may choose between accrual and imputed costs. Accrual costs are only recognised on a cash outlay, imputed costs are assigned to the usage of an input good irrespective of its type of acquisition. Applying these criteria to emission allowances shows that both approaches do not report the same in case of large free allocations like in the first years of the EU ETS. Consequently, the stewardship and decision making functions are found to be better fulfilled when applying an imputed cost accounting.

Financial reporting requirements of EU-based listed firms are determined by IFRS as accounting standards and IFRIC as their interpretations. It took almost two years to pronounce IFRIC 3 (Emission Rights) that was in the end not recommended by the competent EU committees for immediate endorsement into community law. That is why, the interpretation was withdrawn, and the international standard setter IASB as well as the US counterpart FASB are about to adopt a joint standard for emission trading schemes by 2011. For the time being, the interpretation and a US GAAP treatment can be equally applied by European financial accountants. This analysis reviewed both alternatives as well as a further proposal introducing a full fair value measurement. Using a case study approach, especially the different effects on quarterly reports were illustrated, highlighting the technical shortcomings of the withdrawn IFRIC 3. As the US GAAP alternative is set out to map the net effect of excess or supernumerous allowances to accounting income, it conveys a rough signal for stewardship purposes. Decision useful information in a capital market valuation setting is provided in a better fashion by the proposed amendment to IFRIC 3 since its balance sheet items serve as accounting proxy for future cash in- and outflows without generating a mismatch. Even though this normative analysis of accounting treatments employed neoinstitutionalist evaluation criteria, it was unable to estimate the relative usefulness of either alternative. Hence, a third empirical research question was formulated: Which of these treatments gives rise to accounting items according to their economic function? It should be noted that due to the observable information, only the financial accounting and its investor-inclined decision making perspective could be addressed in this question.

Chapter 4 of this study picked up the named three research questions and answered them by furnishing evidence based on several tests using empirical and modelled data. Addressing the aggregate economic consequences, multifactor market models were estimated to isolate the capital market expectation towards the ETS-induced change in firm value. The results presented here reflect the exposure of all major EU power producers during the first trading phase. As an over-compensation for the regulatory costs was found for the complete sample, this corroborates the industrial economics assumption of a monopolistically organised energy sector. Robustness checks showed that firms being subject to end user power price controls or primarily relying on fossil fuels were only able to pass on a part of the regulatory burden. Further, positive valuation effects could also be found for costly input factors like crude oil. It is hence questionable if auctioning emission rights alone can reverse this politically and economically unintended effect of the EU ETS.

Partitioning the sample of power firms according to the prevailing energy market regime (i. e. with market, state or community traditions) and estimating the same multifactor market model gave an insight into differential institutional welfare effects of the EU ETS. This analysis showed that the market arrangements lacked the maturity to mitigate the shift of emission costs, whereas statist regimes provided the most effective buffer. Additional analyses furnished evidence for the assumed transformation processes of European energy markets between 1990 and 2008. A prominent result was that firms operating in traditionally state-based arrangements have witnessed a steep incline in capital market exposure since market liberalisation. This hints at a retirement of the state from its operational responsibility even in those countries having adopted the energy market reforms only recently. In sum, a number of factors was found determining the firm-specific possibility to over-compensate for emission costs like its fuel choice, the degree of regulation or the energy market regime.

As a final test, this analysis employed a value relevance approach to assess the relative usefulness of four financial accounting treatments for emission rights from an investor perspective. Two distinct models are available: price and return regressions. Price regressions can be used to test the pricing properties of ETS-related balance sheet items like allowance assets and emission liabilities. Return regressions help in estimating the properties of the net ETS effect on accounting income. Since a specific disclosure of the respective items is only rarely found in financial reports, it was necessary to model the different accounting alternatives by means of available non-financial information. The general finding was that balance sheet items of all approaches are priced according to their economic function, i. e. assets (liabilities) as value increment (decrement). Earnings showed a differentiated pattern: this study confirmed the distorting effects of the withdrawn IFRIC 3 treatments and at the same time found evidence for a consistent pricing towards the netting US GAAP approach as well as towards a proposed full fair value amendment to IFRIC 3. The inference remained unchanged when controlling for risk factors. As a further robustness check, a novel interacted return model was estimated; it comprised both the incremental effect of emission expenses (vis-à-vis revenues) and the incremental effects of positive and negative allowance returns. Its results again supported the prior findings. Accounting for emission rights offered a nearly unique opportunity to open a "horserace" of competing treatments: comparing the alternatives the empirical section of this study supported the concerns discussed in its normative part towards technical shortcomings of IFRIC 3.

Emission trading is only a recent development in the EU and hence offers broad possibilities for future research. Contributions inspired by institutionalism can further inquire into the opportunistic behaviour of national governments and regulated firms in the run-up to determining the emission budget. Providing empirical results apart from the descriptives presented in this study may support the evidence that exists so far only in anecdotal form. Carbon finance researches will be motivated to investigate into the transitory effects of the future costly allocation under the EU ETS or to find evidence regarding the pricing of US allowances. Normative accounting papers (and empirical ones as soon as sufficient observations are available) can address the properties of the future reporting standard for emission trading schemes. It can be expected that a solution convincing the accounting profession will also be adapted to performance reporting, allowing for further contributions to the management accounting literature.

References

Abadie, Luis M.; Chamorro, José M. (2008): European CO_2 prices and carbon capture investments, in: *Energy Economics*, 30:6, 2992-3015.

Aboody, David; Lev, Baruch (1998): The Value Relevance of Intangibles: The Case of Software Capitalization, in: *Journal of Accounting Research*, 36:Supplement, 161-191.

Åhman, Markus; Burtraw, Dallas; Kruger, Joseph; Zetterberg, Lars (2007): A Ten-Year Rule to guide the allocation of EU emission allowances, in: *Energy Policy*, 35:3, 1718-1730.

Ahmed, Anwer S. (1994): Accounting earnings and future economic rents: An empirical analysis, in: *Journal of Accounting and Economics*, 17:3, 377-400.

Ahmed, Belkaoui (1983): Accrual Accounting and Cash Accounting: Relative Merits of Derived Accounting Indicator Numbers, in: *Journal of Business Finance and Accounting*, 10:2, 299-312.

Ahn, Chang Mo; Thompson, Howard E. (1989): An analysis of some aspects of regulatory risk and the required rate of return for public utilities, in: *Journal of Regulatory Economics*, 1:3, 241-257.

Akerlof, George A. (1970): The Market for "Lemons": Quality Uncertainty and the Market Mechanism, in: *Quarterly Journal of Economics*, 84:3, 488-500.

Alberola, Émilie; Chevallier, Julien (2007): *European carbon prices and banking restrictions: evidence from phase I (2005-2007)*, Working Paper: University of Paris 10.

Alchian, Armen A.; Demsetz, Harold (1972): Production, Information Costs, and Economic Organization, in: *American Economic Review*, 62:5, 777-795.

Alford, Andrew; Jones, Jennifer; Leftwich, Richard; Zmijewski, Mark (1993): The Relative Informativeness of Accounting Disclosures in Different Countries, in: *Journal of Accounting Research*, 31 (Supplement), 183-223.

Allayannis, George; Ofek, Eli (2001): Exchange rate exposure, hedging, and the use of foreign currency derivatives, in: *Journal of International Money and Finance*, 20:2, 273-296.

Amir, Eli; Lev, Baruch (1996): Value-relevance of nonfinancial information: The wireless communications industry, in: *Journal of Accounting and Economics*, 22:1-3, 3-30.

Amundsen, Eirik S.; Bergman, Lars (2006): Why has the Nordic electricity market worked so well?, in: *Utilities Policy*, 14:3, 148-157.

Antheaume, Nicolas (2004): Valuing External Costs – From Theory to Practice: Implications for Full Cost Environmental Accounting, in: *European Accounting Review*, 13:3, 443-464.

Arce, Miguel; Mora, Araceli (2002): Empirical evidence of the effect of European accounting differences on the stock market valuation of earnings and book value, in: *European Accounting Review*, 11:3, 573-599.

Austin, Peter C.; Hoch, Jeffrey S. (2004): Estimating linear regression models in the presence of a censored independent variable, in: *Statistics in Medicine*, 23:3, 411-429.

Averch, Harvey; Leland, L. Johnson (1962): Behavior of the Firm Under Regulatory Constraint, in: *American Economic Review*, 52:5, 1052-1069.

Babiker, Mustafa H.; Criqui, Patrick; Ellerman, A. Denny; Reilly, John M.; Viguier, Laurent L. (2003): Assessing the Impact of Carbon Tax Differentiation in the European Union, in: *Environmental Modeling and Assessment*, 8:3, 187-197.

Bailey, Ian (2007): Market Environmentalism, New Environmental Policy Instruments, and Climate Policy in the United Kingdom and Germany, in: *Annals of the Association of American Geographers*, 97:3, 530-550.

Baiman, Stanley; Demski, Joel S. (1980): Optimal Performance Evaluation and Control Systems, in: *Journal of Accounting Research*, 18 (Studies on Economic Consequences of Financial and Managerial Accounting: Effects on Corporate Incentives and Decisions), 184-220.

Baiman, Stanley; Evans, John Harry (1983): Pre-Decision Information and Participative Management Control Systems, in: *Journal of Accounting Research*, 21:2, 371-395.

Bain, Joe S. (1951): Relation of Profit Rate to Industry Concentration: American Manufacturing, 1936-1940, in: *Quarterly Journal of Economics*, 65:3, 293-324.

Bain, Joe S. (1954): Economies of Scale, Concentration, and the Condition of Entry in Twenty Manufacturing Industries, in: *American Economic Review*, 44:1, 15-39.

Ball, Ray; Brown, Philip (1968): An Empirical Evaluation of Accounting Income Numbers, in: *Journal of Accounting Research*, 6:2, 159-178.

Ball, Ray; Kothari, S. P.; Robin, Ashok (2000): The effect of international institutional factors on properties of accounting earnings, in: *Journal of Accounting and Economics*, 29:1, 1-51.

Ball, Ray; Robin, Ashok; Sadka, Gil (2008): Is financial reporting shaped by equity markets or by debt markets? An international study of timeliness and conservatism, in: *Review of Accounting Studies*, 13:2, 168-205.

Ball, Ray; Shivakumar, Lakshmanan (2005): Earnings quality in UK private firms: comparative loss recognition timeliness, in: *Journal of Accounting and Economics*, 39:1, 83-128.

Barlev, Benzion; Haddad, Joshua Rene (2003): Fair Value Accounting and the Management of the Firm, in: *Critical Perspectives on Accounting*, 14:4, 383-415.

Barnett, A. H. (1980): The Pigouvian Tax Rule Under Monopoly, in: *American Economic Review*, 70:5, 1037-1041.

Baron, David P.; Myerson, Roger B. (1982): Regulating a Monopolist with Unknown Costs, in: *Econometrica*, 50:4, 911-930.

Barr, Nicholas (1992): Economic Theory and the Welfare State: A Survey and Interpretation, in: *Journal of Economic Literature*, 30:2, 741-803.

Barrett, Scott (1998): Political Economy of the Kyoto Protocol, in: *Oxford Review of Economic Policy*, 14:4, 20-39.

Barth, Mary E. (1994): Fair Value Accounting: Evidence from Investment Securities and the Market Valuation of Banks, in: *Accounting Review*, 69:1, 1-25.

Barth, Mary E.; Beaver, William H.; Landsman, Wayne R. (1992): The market valuation implications of net periodic pension cost components, in: *Journal of Accounting and Economics*, 15:1, 27-62.

Barth, Mary E.; Beaver, William H.; Landsman, Wayne R. (2001): The relevance of the value relevance literature for financial accounting standard setting: another view, in: *Journal of Accounting and Economics*, 31:1-3, 77-104.

Barth, Mary E.; Elliott, John A.; Finn, Mark W. (1999): Market Rewards Associated with Patterns of Increasing Earnings, in: *Journal of Accounting Research*, 37:2, 387-413.

Barth, Mary E.; Landsman, Wayne R.; Lang, Mark H. (2008): International Accounting Standards and Accounting Quality, in: *Journal of Accounting Research*, 46:3, 467-498.

Barth, Mary E.; Landsman, Wayne R.; Lang, Mark H.; Williams, Christopher (2009): *Are International Accounting Standards-based and US GAAP-based Accounting Amounts Comparable?*, Working Paper: Stanford University et al.

Barthold, Thomas A. (1994): Issues in the Design of Environmental Excise Taxes, in: *Journal of Economic Perspectives*, 8:1, 133-151.

Bartle, Ian (2002): When Institutions No Longer Matter: Reform of Telecommunications and Electricity in Germany, France and Britain, in: *Journal of Public Policy*, 22:1, 1-27.

Bartov, Eli; Goldberg, Stephen R.; Kim, Myung-Sun (2001): The Valuation-relevance of Earnings and Cash Flows: an International Perspective, in: *Journal of International Financial Management and Accounting*, 12:2, 103-132.

Basu, Sudipta (1997): The conservatism principle and the asymmetric timeliness of earnings, in: *Journal of Accounting and Economics*, 24:1, 3-37.

Basu, Sudipta; Hwang, LeeSeok; Jan, Ching-Lih (2001): *Auditor Conservatism and Quarterly Earnings*, Working Paper Baruch College et al.

Baumol, William J.; Oates, Wallace E. (1975): *The theory of environmental policy: Externalities, public outlays, and the quality of life*, Englewood Cliffs: Prentice-Hall.

Beaver, William H. (1968): The Information Content of Annual Earnings Announcements, in: *Journal of Accounting Research*, 6, 67-92.

Bebbington, Jan; Larrinaga-González, Carlos (2008): Carbon Trading: Accounting and Reporting Issues, in: *European Accounting Review*, 17:4, 697-717.

Becker, Gary S. (1983): A Theory of Competition Among Pressure Groups for Political Influence, in: *Quarterly Journal of Economics*, 98:3, 371-400.

Begley, Joy; Feltham, Gerald A. (1999): An empirical examination of the relation between debt contracts and management incentives, in: *Journal of Accounting and Economics*, 27:2, 229-259.

Beidleman, Carl R. (1973): Income Smoothing: The Role of Management, in: *Accounting Review*, 48:4, 653-667.

Bennedsen, Morten; Feldmann, Sven E. (2002): Lobbying Legislatures, in: *Journal of Political Economy*, 110:4, 919-946.

Benston, George J.; Bromwich, Michael; Wagenhofer, Alfred (2006): Principles- Versus Rules-Based Accounting Standards: The FASB's Standard Setting Strategy, in: *Abacus*, 42:2, 165-188.

Benz, Eva; Trück, Stefan (2009): Modeling the price dynamics of CO_2 emission allowances, in: *Energy Economics*, 31:1, 4-15.

Bergman, Mats A.; Johansson, Per; Bergman, M. A. (2002): Large investments in the pulp and paper industry: a count data regression analysis, in: *Journal of Forest Economics*, 8:1, 29-52.

Bergner, Heinz (1967): Zur Frage des Gutscharakters von Rechten in der Unternehmung, in: *Zeitschrift für betriebswirtschaftliche Forschung*, 19:9, 587-598.

Bernauer, Thomas; Moser, Peter (1996): Reducing Pollution of the River Rhine: The Influence of International Cooperation, in: *Journal of Environment & Development*, 5:4, 389-415.

Betz, Regina; Eichhammer, Wolfgang; Schleich, Joachim (2004): Designing National Allocation Plans For EU-Emissions Trading – A First Analysis of the Outcomes, in: *Energy & Environment*, 15:3, 375-425.

Bhimani, Alnoor; Horngren, Charles T.; Datar, Srikant M.; Foster, George (2008): *Management and Cost Accounting*, 4th ed., Harlow: Pearson Educational.

Biddle, Gary C.; Seow, Gim S. (1991): The Estimation and Determinants of Associations Between Returns and Earnings: Evidence from Cross-industry Comparisons, in: *Journal of Accounting, Auditing & Finance*, 6:2, 183-232.

Biermann, Frank; Dingwerth, Klaus (2004): Global Environmental Change and the Nation State, in: *Global Environmental Politics*, 4:1, 1-22.

Biermann, Frank; Pattberg, Philipp (2008): Global Environmental Governance: Taking Stock, Moving Forward, in: *Annual Review of Environment and Resources*, 33, 277-294.

Biliouri, Daphne (1999): Environmental NGOs in Brussels: How Powerful are Their Lobbying Activities?, in: *Environmental Politics*, 8:2, 173-182.

Binder, John J.; Norton, Seth W. (1999): Regulation, Profit Variability and Beta, in: *Journal of Regulatory Economics*, 15:3, 249-266.

Binkley, Clark S.; Raper, Charles F.; Washburn, Courtland L. (1996): Institutional Ownership of US Timberland: History, Rationale and Implications for Forest Management, in: *Journal of Forestry*, 94:9, 21-28.

Black, Fischer; Jensen, Michael C.; Scholes, Myron (1972): *The Capital Asset Pricing Model: Some Empirical Tests*, Working Paper: Harvard Business School et al.

Blackwell, David (1953): Equivalent Comparisons of Experiments, in: *Annals of Mathematical Statistics*, 24:2, 265-272.

Bleischwitz, Raimund; Fuhrmann, Katrin; Huchler, Elias (2007): *The Sustainability Impact of the EU Emissions Trading System on the European Industry*, Working Paper: College of Europe at Bruges et al.

Blose, Laurence E.; Shieh, Joseph C. P. (1995): The Impact of Gold Price on the Value of Gold Mining Stock, in: *Review of Financial Economics*, 4:2, 125-139.

BMU: Bundesministerium für Umwelt, Naturschutz und Reaktorsicherheit (2007): *Revidierter Nationaler Allokationsplan 2008-2012 für die Bundesrepublik Deutschland.*

BMU: Bundesministerium für Umwelt, Naturschutz und Reaktorsicherheit (2008): *Veräußerung von Emissionsberechtigungen in Deutschland – Jahresbericht 2008.*

Bode, Sven (2006): Multi-period emissions trading in the electricity sector – winners and losers, in: *Energy Policy*, 34:6, 680-691.

Boemare, Catherine; Quirion, Philippe (2002): Implementing greenhouse gas trading in Europe: lessons from economic literature and international experiences, in: *Ecological Economics*, 43:2-3, 213-230.

Börzel, Tanja (2001): Non-compliance in the European Union: pathology or statistical artefact?, in: *Journal of European Public Policy*, 8:5, 803–824.

Boyns, Trevor; Edwards, John Richard (1997): The Construction of Cost Accounting Systems in Britain to 1900: The Case of the Coal, Iron and Steel Industries, in: *Business History*, 39:3, 1-29.

Bradley, Ian; Price, Catherine (1988): The Economic Regulation of Private Industries by Price Constraints, in: *Journal of Industrial Economics*, 37:1, 99-106.

Braun, Marcel (2009): The evolution of emissions trading in the European Union – The role of policy networks, knowledge and policy entrepreneurs, in: *Accounting, Organizations and Society*, 34:3/4, 469-487.

Brown, Stephen; Lo, Kin; Lys, Thomas (1999): Use of R^2 in accounting research: measuring changes in value relevance over the last four decades, in: *Journal of Accounting and Economics*, 28:2, 83-115.

Bruno, Michael; Sachs, Jeffrey D. (1985): *Economics of Worldwide Stagflation*, Oxford: Blackwell.

Buchanan, James M. (1954): Social Choice, Democracy, and Free Markets, in: *Journal of Political Economy*, 62:2, 114-123.

Buchanan, James M. (1999): *The collected work of James M. Buchanan*, Indianapolis: Liberty Fund.

Bundeskartellamt (2006): *Sachstandspapier zur Vorbereitung der mündlichen Verhandlung in Sachen Emissionshandel und Strompreisbildung*.

Bunyard, Peter (1990): Uranium: do leave it in the ground!, in: *Energy Policy*, 18:8, 689-691.

Burgi, Martin; Lange, Knut Werner (2006): Der Emissionshandel als Herausforderung für die gesamte Rechtsordnung, in: *Zeitschrift für das gesamte Handelsrecht und Wirtschaftsrecht*, 170:5, 539-566.

Burtraw, Dallas; Evans, David A.; Krupnick, Alan; Palmer, Karen; Toth, Russel (2005): Economics of Pollution Trading for SO_2 and NO_x, in: *Annual Review of Environment and Resources*, 30, 253-289.

Burtraw, Dallas; Palmer, Karen; Bharvirkar, Ranjit; Paul, Anthony (2002): The Effect on Asset Values of the Allocation of Carbon Dioxide Emission Allowances, in: *Electricity Journal*, 15:5, 51-62.

Busse von Colbe, Walther (1989): Netto-Substanzerhaltung als integratives Konzept für Bilanzierung und Kostenrechnung, in: Delfmann, Werner (Ed.): *Der Integrationsgedanke in der Betriebswirtschaftslehre: Helmut Koch zum 70. Geburtstag*, Wiesbaden: Gabler, 73-85.

Cahan, Steven F.; Emanuel, David; Sun, Jerry (2009): The effect of earnings quality and country-level institutions on the value relevance of earnings, in: *Review of Quantitative Finance & Accounting*, 33:4, 371-391.

Callon, Michel (2009): Civilizing markets: Carbon trading between *in vitro* and *in vivo* experiments, in: *Accounting, Organizations and Society*, 34:3/4, 535-548.

Cansier, Dieter; Krumm, Raimund (1997): Air pollutant taxation: an empirical survey, in: *Ecological Economics*, 23:1, 59-70.

Carlson, Curtis; Burtraw, Dallas; Cropper, Maureen L.; Palmer, Karen L. (2000): Sulfur Dioxide Control by Electric Utilities: What Are the Gains from Trade?, in: *Journal of Political Economy*, 108:6, 1292-1326.

Carlton, Dennis W.; Loury, Glenn C. (1980): The Limitations of Pigouvian Taxes as a Long-Run Remedy for Externalities, in: *Quarterly Journal of Economics*, 95:3, 559-566.

CEER: Council of European Energy Regulators (2005): *CEER Regulatory Benchmark Report 2005.*

CEPI: Confederation of European Paper Industries (2006): *Paper & Board Market Shares of the Top 10 Companies by Region – 2005.*

Chen, Andrew H.; Merville, Larry J. (1986): An Analysis of Divestiture Effects Resulting from Deregulation, in: *Journal of Finance*, 41:5, 997-1010.

Chenhall, Robert H.; Langfield-Smith, Kim (1998): Adoption and benefits of management accounting practices: an Australian study, in: *Management Accounting Research*, 9:1, 1-19.

Chiapello, Eve; Medjad, Karim (2009): An unprecedented privatisation of mandatory standard-setting: The case of European accounting policy, in: *Critical Perspectives on Accounting*, 20:4, 448-468.

Childs, Paul D.; Ott, Steven H.; Riddiough, Timothy J. (2001): Valuation and Information Acquisition Policy for Claims Written on Noisy Real Assets, in: *Financial Management*, 30:2, 45-75.

Chmielewicz, Klaus (1970): Wirtschaftsgut, in: Kosiol, Erich (Ed.): *Enzyklopädie der Betriebswirtschaftslehre*, Stuttgart: Schäffer-Poeschel, cols. 1943-1952.

Christensen, John A.; Demski, Joel S. (1995): Project selection and audited accrual measurement in a multi-task setting, in: *European Accounting Review*, 4:3, 405-432.

Christensen, John A.; Demski, Joel S. (2003): *Accounting Theory: An Information Content Perspective*, Boston et al.: McGraw-Hill/Irwin.

Clarke, Roger G. (1980): The Effect of Fuel Adjustment Clauses on the Systematic Risk and Market Values of Electric Utilities, in: *Journal of Finance*, 35:2, 347-358.

Coase, Ronald H. (1960): The Problem of Social Cost, in: *Journal of Law and Economics*, 1:3, 1-44.

Collins, Daniel W.; Kothari, S. P. (1989): An analysis of intertemporal and cross-sectional determinants of earnings response coefficients, in: *Journal of Accounting and Economics*, 11:2-3, 143-181.

Collins, Daniel W.; Maydew, Edward L.; Weiss, Ira S. (1997): Changes in the value-relevance of earnings and book values over the past forty years, in: *Journal of Accounting and Economics*, 24:1, 39-67.

Colpier, Ulrika Claeson; Cornland, Deborah (2002): The economics of the combined cycle gas turbine – an experience curve analysis, in: *Energy Policy*, 30:4, 309-316.

Competition Commission (2006): *Domestic bulk liquefied petroleum gas*, Final Report.

Convery, Frank J.; De Perthuis, Christian; Ellerman, Denny (2008): *The European Carbon Market In Action: Lessons From The First Trading Period*, Working Paper: MIT Center for Energy and Environmental Policy Research.

Convery, Frank J.; Redmond, Luke (2007): Market and Price Developments in the European Union Emissions Trading Scheme, in: *Review of Environmental Economics and Policy*, 1:1, 88-111.

Cook, Allan (2009): Emission rights: From costless activity to market operations, in: *Accounting, Organizations and Society*, 34:3-4, 456-468.

Core, John E.; Guay, Wayne R.; Van Buskirk, Andrew (2003): Market valuations in the New Economy: an investigation of what has changed, in: *Journal of Accounting and Economics*, 34:1-3, 43-67.

Council of the EU (2009): *Press Release 8434/09: Council adopts climate-energy legilsative package.*

Crals, Evy; Vereeck, Lode (2005): Taxes, Tradable Rights and Transaction Costs, in: *European Journal of Law and Economics*, 20:2, 199-223.

Crocker, Thomas D. (1966): The Structuring of Atmospheric Pollution Control Systems, in: Wolozin, Harold (Ed.): *The Economics of Air Pollution: A Symposium*, 2nd ed., New York: Norton.

Cropper, Maureen L.; Oates, Wallace E. (1992): Environmental Economics: A Survey, in: *Journal of Economic Literature*, 30:2, 675-740.

Cullenberg, Stephen (1999): Overdetermination, Totality, and Institutions: A Genealogy of a Marxist Institutionalist Economics, in: *Journal of Economic Issues*, 33:4, 801-815.

Dal Bó, Ernesto (2006): Regulatory Capture: A Review, in: *Oxford Review of Economic Policy*, 22:2, 202-225.

Dales, John Harkness (1968): Land, Water, and Ownership, in: *Canadian Journal of Economics/Revue canadienne d'Économique*, 1:4, 797-804.

Dales, John Harkness (1970): *Pollution, property & prices: An essay in policy-making and economics*, 2nd ed., Toronto: University of Toronto Press.

Dalton, James A.; Penn, David W. (1976): The Concentration-Profitability Relationship: Is There a Critical Concentration Ratio?, in: *Journal of Industrial Economics*, 25:2, 133-142.

Daniels, M. B. (1933): The Valuation of Fixed Assets, in: *Accounting Review*, 8:4, 302-316.

Daskalakis, George; Markellos, Raphael N. (2007): *Are the European Carbon Markets Efficient?*, Working Paper: Athens University.

Davidson, Wallace N.; Rangan, Nanda; Rosenstein, Stuart (1997): Regulation and Systematic Risk in the Electric Utility Industry: A Test of the Buffering Hypothesis, in: *Financial Review*, 32:1, 163-184.

de B. Harris, Frederick H. (1990): Economic Negligence, Moral Hazard, and the Coase Theorem, in: *Southern Economic Journal*, 56:3, 698.

de Jong, Abe; Ligterink, Jeroen; Macrae, Victor (2006): A Firm-Specific Analysis of the Exchange-Rate Exposure of Dutch Firms, in: *Journal of International Financial Management and Accounting*, 17:1, 1-28.

DEHSt: Deutsche Emissionshandelsstelle (2004): *Anwendung und Berechnung der anteiligen Kürzung der Zuteilungsmengen für die erste Zuteilungsperiode nach § 4 Absatz 4 ZuG 2007.*

DEHSt: Deutsche Emissionshandelsstelle (2005): *Implementation of emissions trading in the EU: National Allocation Plans of all EU states: Brief fact sheets of EU member state allocation plans.*

Demailly, Damien; Quirion, Philippe (2006): CO_2 abatement, competitiveness and leakage in the European cement industry under the EU ETS: grandfathering versus output-based allocation, in: *Climate Policy*, 6:1, 93-113.

Demailly, Damien; Quirion, Philippe (2008): European Emission Trading Scheme and competitiveness: A case study on the iron and steel industry, in: *Energy Economics*, 30:4, 2009-2027.

Demsetz, Harold (1968): Why Regulate Utilities?, in: *Journal of Law and Economics*, 11:1, 55-65.

Demski, Joel S.; Feltham, Gerald A. (1976): *Cost Determination: A Conceptual Approach*, Ames: Iowa State University Press.

Demski, Joel S.; Sappington, David E. M. (1990): Fully Revealing Income Measurement, in: *Accounting Review*, 65:2, 365-383.

Dewees, D. N.; Sims, W. A. (1976): The Symmetry of Effluent Charges and Subsidies for Pollution Control, in: *Canadian Journal of Economics/Revue canadienne d'Économique*, 9:2, 323-331.

Dickhaut, John W.; Lere, John C. (1983): Comparison of Accounting Systems and Heuristics in Selecting Economic Optima, in: *Journal of Accounting Research*, 21:2, 495-513.

DiMaggio, Paul J.; Powell, Walter W. (1983): The Iron Cage Revisited: Institutional Isomorphism and Collective Rationality in Organizational Fields, in: *American Sociological Review*, 48:2, 147-160.

Dissou, Yazid (2005): Cost-effectiveness of the performance standard system to reduce CO2 emissions in Canada: a general equilibrium analysis, in: *Resource and Energy Economics*, 27:3, 187-207.

Dixit, Avinash; Olson, Mancur (2000): Does voluntary participation undermine the Coase Theorem?, in: *Journal of Public Economics*, 76:3, 309-335.

DMO: UK Debt Management Office (2009): *EU Emissions Trading Scheme: Auction Information.*

Dnes, Antony W.; Seaton, Jonathan S. (1999): The regulation of electricity: results from an event study, in: *Applied Economics*, 31:5, 609-618.

Domanico, Fabio (2007): Concentration in the European electricity industry: The internal market as solution, in: *Energy Policy*, 35:10, 5064-5076.

Douglas, Ian; Hodgson, Rob; Lawson, Nigel (2002): Industry, environment and health through 200 years in Manchester, in: *Ecological Economics*, 41:2, 235-255.

Dumanski, Julian (2004): Carbon Sequestration, Soil Conservation, and the Kyoto Protocol: Summary of Implications, in: *Climatic Change*, 65:3, 255-261.

Dye, Ronald A. (2004): Strategy Selection and Performance Measurement Choice When Profit Drivers Are Uncertain, in: *Management Science*, 50:12, 1624-1637.

Easton, Peter D.; Harris, Trevor S. (1991): Earnings As an Explanatory Variable for Returns, in: *Journal of Accounting Research*, 29:1, 19-36.

Easton, Peter D.; Sommers, Gregory A. (2003): Scale and the Scale Effect in Market-based Accounting Research, in: *Journal of Business Finance & Accounting*, 30:1/2, 25-55.

Eberlein, Burkhard (2000): Institutional Change and Continuity in German Infrastructure Management: The Case of Electricity Reform, in: *German Politics*, 9:3, 81-104.

Edwards, John Richard; Boyns, Trevor; Matthews, Mark (2003): Costing, pricing and politics in the British steel industry, 1918–1967, in: *Management Accounting Research*, 14:1, 25-49.

EFRAG: European Financial Reporting Advisory Group (2003): *Comment Letter regarding IFRIC Draft Interpretation 1 Emission Rights*.

Ehni, Carmen; Fink, Simon; Jäger, Alexander; Thiem, Janina (2004): Privatisierungsverläufe in drei Infrastruktursektoren im internationalen Vergleich, in: Schneider, Volker; Tenbücken, Marc (Eds.): *Der Staat auf dem Rückzug: Die Privatisierung öffentlicher Infrastrukturen*, Frankfurt et al.: Campus, 115-153.

Ehrt, Robert (1967): *Die Zurechenbarkeit von Kosten auf Leistungen: auf der Grundlage kausaler und finaler Beziehungen*, Stuttgart et al.: Kohlhammer, Dissertation, University of Mannheim (Wirtschaftshochschule).

El-Fadel, M.; Zeinati, M.; Ghaddar, N.; Mezher, T. (2001): Uncertainty in estimating and mitigating industrial related GHG emissions, in: *Energy Policy*, 29:12, 1031-1043.

El-Sharif, Idris; Brown, Dick; Burton, Bruce; Nixon, Bill; Russell, Alex (2005): Evidence on the nature and extent of the relationship between oil prices and equity values in the UK, in: *Energy Economics*, 27:6, 819-830.

Ellerman, Denny; Buchner, Barbara K. (2008): Over-Allocation or Abatement? A Preliminary Analysis of the EU ETS Based on the 2005-06 Emissions Data, in: *Environmental and Resource Economics*, 41:2, 267-287.

Endres, Alfred; Ohl, Cornelia (2005): Kyoto, Europe? – An Economic Evaluation of the European Emission Trading Directive, in: *European Journal of Law and Economics*, 19:1, 17-39.

Endres, Walter (1976): Menschen und Gegenstände im Betrieb und seiner Umwelt, in: *Zeitschrift für betriebswirtschaftliche Forschung*, 28:12, 781-804.

ERGEG: European Regulators' Group for Electricity and Gas (2007): *End-user energy price regulation.*

European Commission (2005): *Report on Progress in Creating the Internal Gas and Electricity Market: Technical Annex.*

European Commission (2007): *Accompanying document to the Communication from the Commission to the Council and the European Parliament: Prospects for the internal gas and electricity market: Implementation report.*

European Commission (2008): *Accompanying document to the Report on Progress in Creating the Internal Gas and Electricity Market.*

European Commission (2009): *Emission Trading System (EU ETS): Auctioning.*

Ewert, Ralf; Wagenhofer, Alfred (2000): *Interne Unternehmensrechnung*, Berlin et al.: Springer.

Faff, Robert W.; Brailsford, Timothy J. (1999): Oil price risk and the Australian stock market, in: *Journal of Energy Finance and Development*, 4:3, 69-87.

Faff, Robert W.; Chan, Howard (1998): A multifactor model of gold industry stock returns: evidence from the Australian equity market, in: *Applied Financial Economics*, 8:1, 21-28.

Faff, Robert W.; Hillier, David (2004): An International Investigation of the Factors that Determine Conditional Gold Betas, in: *Financial Review*, 39:3, 473-488.

Fama, Eugene F. (1991): Efficient Capital Markets: II, in: *Journal of Finance*, 46:5, 1575-1617.

Fama, Eugene F.; French, Kenneth R. (1995): Size and Book-to-Market Factors in Earnings and Returns, in: *Journal of Finance*, 50:1, 131-155.

Fama, Eugene F.; French, Kenneth R. (1997): Industry costs of equity, in: *Journal of Financial Economics*, 42:2, 153-193.

Fama, Eugene F.; French, Kenneth R. (2004): The Capital Asset Pricing Model: Theory and Evidence, in: *Journal of Economic Perspectives*, 18:3, 25-46.

Farrell, Joseph (1987): Information and the Coase Theorem, in: *Journal of Economic Perspectives*, 1:2, 113-129.

Felder, Stefan; Rutherford, Thomas F. (1993): Unilateral CO2 Reductions and Carbon Leakage: The Consequences of International Trade in Oil and Basic Materials, in: *Journal of Environmental Economics and Management*, 25:2, 162-176.

Feltham, Gerald A. (1968): The Value of Information, in: *Accounting Review*, 43:4, 684-696.

Feltham, Gerald A.; Ohlson, James A. (1995): Valuation and Clean Surplus Accounting for Operating and Financial Activities, in: *Contemporary Accounting Research*, 11:2, 689-731.

Ferrari, Alessandra; Giulietti, Monica (2005): Competition in electricity markets: international experience and the case of Italy, in: *Utilities Policy*, 13:3, 247-255.

Franciosi, Robert; Isaac, R. Mark; Pingry, David E.; Reynolds, Stanley S. (1993): An Experimental Investigation of the Hahn-Noll Revenue Neutral Auction for Emissions Licenses, in: *Journal of Environmental Economics and Management*, 24:1, 1-24.

Francis, Jennifer; Schipper, Katherine (1999): Have Financial Statements Lost Their Relevance?, in: *Journal of Accounting Research*, 37:2, 319-352.

Freeman, Robert N.; Tse, Senyo Y. (1992): A Nonlinear Model of Security Price Responses to Unexpected Earnings, in: *Journal of Accounting Research*, 30:2, 185-209.

Frese, Erich; Kloock, Josef (1989): Internes Rechnungswesen und Organisation aus der Sicht des Umweltschutzes, in: *Betriebswirtschaftliche Forschung und Praxis*, 41:1, 1-29.

Friedman, Laurence A.; Neumann, Bruce R. (1980): The Effects of Opportunity Costs on Project Investment Decisions: A Replication and Extension, in: *Journal of Accounting Research*, 18:2, 407-419.

Garber, Steven; Hammitt, James K. (1998): Risk Premiums for Environmental Liability: Does Superfund Increase the Cost of Capital?, in: *Journal of Environmental Economics and Management*, 36:3, 267-294.

Geithman, Frederick E.; Marvel, Howard P.; Weiss, Leonard W. (1981): Concentration, Price, and Critical Concentration Ratios, in: *Review of Economics and Statistics*, 63:3, 346-353.

Genschel, Philipp (2005): Globalization and the transformation of the tax state, in: Leibfried, Stephan; Zürn, Michael (Eds.): *Transformations of the State?*, Cambridge (UK): Cambridge University Press, 53-71.

Georg, Susse; Røpke, Inge; Jørgensen, Ulrik (1992): Clean technology — Innovation and environmental regulation, in: *Environmental and Resource Economics*, 2:6, 533-550.

Gibson, Kathy (1996): The Problem with Reporting Pollution Allowances: Reporting is not the Problem, in: *Critical Perspectives on Accounting*, 7:6, 655-665.

Glaeser, Edward; Johnson, Simon; Shleifer, Andrei (2001): Coase versus the Coasians, in: *Quarterly Journal of Economics*, 116:3, 853-899.

Glaum, Martin; Klöckner, André (2009): Hedge Accounting nach IAS 39 in der Praxis, in: *Zeitschrift für internationale und kapitalmarktorientierte Rechnungslegung*, 9:6, 329-340.

Goulder, Lawrence H.; Parry, Ian W. H. (2008): Instrument Choice in Environmental Policy, in: *Review of Environmental Economics and Policy*, 2:2, 152-174.

Goulder, Lawrence H.; Parry, Ian W. H.; Williams, Roberton C.; Burtraw, Dallas (1999): The cost-effectiveness of alternative instruments for environmental protection in a second-best setting, in: *Journal of Public Economics*, 72:3, 329-360.

Göx, Robert F.; Wagenhofer, Alfred (2009): Optimal impairment rules, in: *Journal of Accounting and Economics*, 48:1, 2-16.

Griffin, John M.; Stulz, René M. (2001): International Competition and Exchange Rate Shocks: A Cross-Country Industry Analysis of Stock Returns, in: *Review of Financial Studies*, 14:1, 215-241.

Grout, Paul A.; Zalewska, Anna (2006): The impact of regulation on market risk, in: *Journal of Financial Economics*, 80:1, 149-184.

Grubb, Michael; Neuhoff, Karsten (2006): Allocation and competitiveness in the EU emission trading scheme: policy overview, in: *Climate Policy*, 6:1, 7-30.

Günther, Edeltraud (1994): *Ökologieorientiertes Controlling: Konzeption eines Systems zur ökologieorientierten Steuerung und empirische Validierung*, München: Vahlen, Dissertation, University of Augsburg.

Günther, Edeltraud (2006): Accounting for emission rights, in: Antes, Ralf; Hansjürgens, Bernd; Letmathe, Peter (Eds.): *Emissions Trading and Business*, Heidelberg: Physica-Verlag, 219-239.

Gutenberg, Erich (1958): *Einführung in die Betriebswirtschaftslehre*, unchanged reprint in 1975, Wiesbaden: Gabler.

Häge, Frank M.; Schneider, Volker (2004): Hauptachsen staatlicher Redimensionierung: Die Rolle von Europäisierung und Globalisierung, in: Schneider, Volker; Tenbücken, Marc (Eds.): *Der Staat auf dem Rückzug: Die Privatisierung öffentlicher Infrastrukturen*, Frankfurt et al.: Campus, 280-313.

Hahn, Robert W. (1984): Market Power and Transferable Property Rights, in: *Quarterly Journal of Economics*, 99:4, 753-765.

Hahn, Robert W.; Stavins, Robert N. (1992): Economic Incentives for Environmental Protection: Integrating Theory and Practice, in: *American Economic Review*, 82:2, 464-468.

Hancher, Leigh; de Hauteclocque, Adrien (2010): *Manufacturing the EU Energy Markets: The Current Dynamics of Regulatory Practice*, Working Paper: European University Institute.

Hantke-Domas, Michael (2003): The Public Interest Theory of Regulation: Non-Existence or Misinterpretation?, in: *European Journal of Law and Economics*, 15:2, 165-194.

Harris, Trevor S.; Lang, Mark; Möller, Hans Peter (1994): The Value Relevance of German Accounting Measures: An Empirical Analysis, in: *Journal of Accounting Research*, 32:2, 187-209.

Hax, Herbert (1989): Investitionsrechnung und Periodenerfolgsmessung, in: Delfmann, Werner (Ed.): *Der Integrationsgedanke in der Betriebswirtschaftslehre*, Wiesbaden: Gabler, 153-170.

He, Jia; Ng, Lilian K. (1998): The Foreign Exchange Exposure of Japanese Multinational Corporations, in: *Journal of Finance*, 53:2, 733-753.

Healy, Paul M.; Myers, Steward C.; Howe, Christopher D. (2002): R&D Accounting and the Tradeoff Between Relevance and Objectivity, in: *Journal of Accounting Research*, 40:3, 677-710.

Heinen, Edmund (1983): *Betriebswirtschaftliche Kostenlehre: Kostentheorie und Kostenentscheidungen*, 6th ed., Wiesbaden: Gabler.

Helm, Dieter (2002): Energy policy: security of supply, sustainability and competition, in: *Energy Policy*, 30:3, 173-184.

Helm, Dieter; Pearce, David (1990): Assessment: Economic Policy towards the Environment, in: *Oxford Review of Economic Policy*, 6:1, 1-16.

Hermes, Oliver; Jödicke, Ralf (2004): Bilanzierung von Emissionsrechten nach IFRS, in: *Zeitschrift für internationale und kapitalmarktorientierte Rechnungslegung*, 4:7-8, 287-298.

Hesford, James W.; Lee, Sung-Han; van der Stede, Wim A.; Young, S. Mark (2006): Management Accounting: A Bibliographic Study, in: *Handbooks of Management Accounting Research*, 1, 3-26.

Hobsbawm, Eric (1995): *Das Zeitalter der Extreme: Weltgeschichte des 20. Jahrhunderts*, Munich et al.: Hanser.

Hollingsworth, J. Rogers (2000): Doing institutional analysis: implications for the study of innovations, in: *Review of International Political Economy*, 7:4, 595-644.

Holmström, Bengt (1979): Moral Hazard and Observability, in: *Bell Journal of Economics*, 10:1, 74-91.

Holthausen, Robert W.; Watts, Ross L. (2001): The relevance of the value-relevance literature for financial accounting standard setting, in: *Journal of Accounting and Economics*, 31:1-3, 3-75.

Holzer, H. Peter; Norreklit, Hanne (1991): Some thoughts on cost accounting developments in the United States, in: *Management Accounting Research*, 2:1, 3-13.

Hommel, Michael; Wolf, Sandra (2005): IFRIC 3: Bilanzierung von Emissionsrechten nach IFRS – mehr Schadstoffe im Abschluss, in: *Betriebs-Berater*, 60:6, 315-321.

Hughes, K. E. (2000): The Value Relevance of Nonfinancial Measures of Air Pollution in the Electric Industry, in: *Accounting Review*, 75:2, 209-228.

Hughes, K. E.; Reynolds, Kenneth (2001): Uncertainty Associated with Future Environmental Costs and the Market's Differential Response to Earnings Information, in: *Journal of Business Finance & Accounting*, 28:9-10, 1351-1386.

Humphreys, Peter; Padgett, Stephen (2006): Globalization, the European Union, and Domestic Governance in Telecoms and Electricity, in: *Governance: An International Journal of Policy, Administration and Institutions*, 19:3, 383-406.

IASB: International Accounting Standards Board (2005): *Information for Observers: IFRIC Meeting 31 March-1 April 2005*.

IASB: International Accounting Standards Board (2008): *Project Update: Emission Trading Schemes*.

IASB: International Accounting Standards Board (2009): *Information for Observers: Board Meeting March 2009*.

IPCC: Intergovernmental Panel on Climate Change (2001): *Third Assessment Report: Climate Change 2001: The Scientific Basis (Summary Document)*.

IPCC: Intergovernmental Panel on Climate Change (2007): *Fourth Assessment Report: Climate Change 2007: The Physical Science Basis (Technical Summary)*.

Jabko, Nicolas (2006): *Playing the market: a political strategy for uniting Europe, 1985-2005*, Ithaca: Cornell University Press.

Jaffe, Adam B.; Stavins, Robert N. (1995): Dynamic Incentives of Environmental Regulations: The Effects of Altenative Policy Instruments on Technology Diffusion, in: *Journal of Environmental Economics and Management*, 29 (Supplement), S-43-S-63.

Jamasb, Tooraj; Pollitt, Michael (2000): Benchmarking and regulation: international electricity experience, in: *Utilities Policy*, 9:3, 107-130.

Jamasb, Tooraj; Pollitt, Michael (2005): *Electricity Market Reform in the European Union: Review of Progress toward Liberalization & Integration*, Working Paper: MIT Center for Energy and Environmental Policy Research.

James, Oliver (2000): Regulation inside Government: Public Interest Justifications and Regulatory Failures, in: *Public Administration*, 78:2, 327-343.

Jarrell, Gregg A. (1978): The Demand for State Regulation of the Electric Utility Industry, in: *Journal of Law and Economics*, 21:2, 269-295.

Jenkins, David S.; Kane, Gregory D.; Velury, Uma (2009): Earnings Conservatism and Value Relevance Across the Business Cycle, in: *Journal of Business Finance & Accounting*, 36:9/10, 1041-1058.

Jensen, Michael C. (1968): The Performance of Mutual Funds in the Period 1945-1964, in: *Journal of Finance*, 23:2, 389-416.

Jensen, Michael C.; Meckling, William H. (1976): Theory of the firm: Managerial behavior, agency costs and ownership structure, in: *Journal of Financial Economics*, 3:4, 305-360.

Johnston, Derek M.; Sefcik, Stephan E.; Soderstrom, Naomi S. (2008): The Value Relevance of Greenhouse Gas Emissions Allowances: An Exploratory Study in the Related United States SO_2 Market, in: *European Accounting Review*, 17:4, 747-764.

Jones, Jefferson P.; Morton, Richard M.; Schaefer, Thomas F. (2000): Valuation Implications of Investment Opportunities and Earnings Performance, in: *Review of Quantitative Finance and Accounting*, 15:1, 21-35.

Jordan, Andrew; Wurzel, Rüdiger K. W.; Zito, Anthony (2003): Comparative Conclusions – 'New' Environmental Policy Instruments: An Evolution or a Revolution in Environmental Policy?, in: *Environmental Politics*, 12:1, 201-224.

Jorion, Philippe (1990): The Exchange-Rate Exposure of U.S. Multinationals, in: *Journal of Business*, 63:3, 331-345.

Joskow, Paul L.; Noll, Roger C. (1981): Regulation in Theory and Practice: An Overview, in: Fromm, Gary (Ed.): *Studies in Public Regulation*, Cambridge (US): MIT Press, 1-78.

Joskow, Paul L.; Schmalensee, Richard (1998): The Political Economy of Market-Based Environmental Policy: The U.S. Acid Rain Program, in: *Journal of Law and Economics*, 41:1, 37-83.

Käfer, Karl (1964): *Standardkostenrechnung*, 2nd ed., Stuttgart: Poeschel.

Kahn, Shulamit; Knittel, Christopher R. (2003): *The Impact of the Clean Air Act Amendments of 1990 on Electric Utilities and Coal Mines: Evidence from the Stock Market*, Center for the Study of Energy Markets.

Kang, Sok-Hyon; Zhao, Yuping (2010): Information Content and Value Relevance of Depreciation: A Cross-Industry Analysis, in: *Accounting Review*, 85:1, 227-260.

Kaplan, Robert S. (1984): The Evolution of Management Accounting, in: *Accounting Review*, 59:3, 390-418.

Kilger, Wolfgang; Pampel, Jochen; Vikas, Kurt (2002): *Flexible Plankostenrechnung und Deckungsbeitragsrechnung*, 11th ed., Wiesbaden: Gabler.

Kirchner, Paul (1961): Theory and Research in Management Accounting, in: *Accounting Review*, 36:1, 43-49.

Kirkulak, Berna; Balsari, Cağnur Kaytmaz (2009): Value Relevance of Inflation-adjusted Equity and Income, in: *International Journal of Accounting*, 44:4, 363-377.

Kloock, Josef (1993): Neuere Entwicklungen betrieblicher Umweltkostenrechnungen, in: Wagner, Gerd Rainer (Ed.): *Betriebswirtschaft und Umweltschutz*, Stuttgart: Schäffer-Poeschel, 179-206.

Koch, Helmut (1958): Zur Diskussion über den Kostenbegriff, in: *Zeitschrift für handelswissenschaftliche Forschung*, 10:7/8, 355-399.

Koch, Helmut (1966): *Grundprobleme der Kostenrechnung*, Cologne et al.: Westdeutscher Verlag.

Kolk, Ans; Levy, David; Pinkse, Jonatan (2008): Corporate Responses in an Emerging Climate Regime: The Institutionalization and Commensuration of Carbon Disclosure, in: *European Accounting Review*, 17:4, 719-745.

Kolstad, Charles D. (1986): Empirical Properties of Economic Incentives and Command-and-Control Regulations for Air Pollution Control, in: *Land Economics*, 62:3, 250-268.

Kosiol, Erich (1953): *Kalkulatorische Buchhaltung (Betriebsbuchhaltung): Systematische Darstellung der Betriebsabrechnung und der kurzfristigen Erfolgsrechnung*, 5th ed., Wiesbaden: Gabler.

Kosiol, Erich (1958): Kritische Analyse der Wesensmerkmale des Kostenbegriffs, in: Kosiol, Erich; Schlieper, Friedrich (Eds.): *Betriebsökonomisierung durch Kostenanalyse, Absatzrationalisierung und Nachwuchserziehung: Festschrift für Professor Dr. Dr. h. c. Rudolf Seÿffert zu seinem 65. Geburtstag*, Cologne et al.: Westdeutscher Verlag, 7-37.

Kosiol, Erich (1975): *Die Unternehmung als wirtschaftliches Aktionszentrum*, 3rd ed., Reinbek: Rowohlt.

Kothari, S. P. (2001): Capital markets research in accounting, in: *Journal of Accounting and Economics*, 31:1-3, 105-231.

Kothari, S. P.; Shanken, Jay (2003): Time-series coefficient variation in value-relevance regressions: a discussion of Core, Guay, and Van Buskirk and new evidence, in: *Journal of Accounting and Economics*, 34:1-3, 69-87.

Kothari, S. P.; Zimmerman, Jerold L. (1995): Price and return models, in: *Journal of Accounting and Economics*, 20:2, 155-192.

Kousenidis, Dimitrios V.; Ladas, Anestis C. ; Negakis, Christos I. (2009): Value relevance of conservative and non-conservative accounting information, in: *International Journal of Accounting*, 44:3, 219-238.

Kruger, Joseph; Pizer, William A. (2004): *The EU Emissions Trading Directive: Opportunities and Potential Pitfalls*, Working Paper: Resources for the Future, Washington, D.C.

Küpper, Hans-Ulrich; Mattessich, Richard (2005): Twentieth century accounting research in the German language area, in: *Accounting, Business & Financial History*, 15:3, 345-410.

Kwon, Young K. (1989): Accrual Versus Cash-Basis Accounting Methods: An Agency-Theoretic Comparison, in: *Journal of Accounting and Public Policy*, 8:4, 267-281.

Labatt, Sonia; White, Rodney R. (2007): *Carbon Finance: The Financial Implications of Climate Change*, Hoboken: Wiley.

Laffont, Jean-Jacques; Tirole, Jean (1991): The Politics of Government Decision-Making: A Theory of Regulatory Capture, in: *Quarterly Journal of Economics*, 106:4, 1089-1127.

Lange, Christoph; Fischer, Regina (1998): Umweltschutzbezogene Kostenrechnung auf Basis der Einzelkosten- und Deckungsbeitragsrechnung als Instrument des Controlling, in: *Zeitschrift für Betriebswirtschaft*, 68 (Ergänzungsheft 1), 107-123.

Larsen, Anders; Pedersen, Lene Holm; Sørensen, Eva Moll; Olsen, Ole Jess (2006): Independent regulatory authorities in European electricity markets, in: *Energy Policy*, 34:17, 2858-2870.

Laux, Helmut (2006): *Unternehmensrechnung, Anreiz und Kontrolle: Die Messung, Zurechnung und Steuerung des Erfolges als Grundprobleme der Betriebswirtschaftslehre*, 3rd ed., Berlin et al.: Springer.

Lee, Lucy C.; Bedford, Norton M. (1969): An Information Theory Analysis of the Accounting Process, in: *Accounting Review*, 44:2, 256-275.

Lehman, Glen (1996): Environmental Accounting: Pollution Permits or Selling the Environment, in: *Critical Perspectives on Accounting*, 7:6, 667-676.

Lehmann, Max Rudolf (1956): *Allgemeine Betriebswirtschaftslehre: Allgemeine Theorie der Betriebswirtschaft*, 3rd ed., Wiesbaden: Gabler.

Leibfried, Stephan; Zürn, Michael (2005): Reconfiguring the national constellation, in: Leibfried, Stephan; Zürn, Michael (Eds.): *Transformations of the State?*, Cambridge (UK): Cambridge University Press, 1-36.

Lejano, Raul P.; Hirose, Rei (2005): Testing the assumptions behind emissions trading in non-market goods: the RECLAIM program in Southern California, in: *Environmental Science & Policy*, 8:4, 367-377.

Leland, Hayne E. (1972): Theory of the Firm Facing Uncertain Demand, in: *American Economic Review*, 62:3, 278-291.

Letmathe, Peter; Doost, Roger K. (2000): Environmental cost accounting and auditing, in: *Managerial Auditing Journal*, 15:8, 424.430.

Lev, Baruch; Sougiannis, Theodore (1996): The capitalization, amortization, and value-relevance of R&D, in: *Journal of Accounting and Economics*, 21:1, 107-138.

Lewis, Tracy R. (1996): Protecting the environment when costs and benefits are privately known, in: *RAND Journal of Economics*, 27:4, 819-847.

Liebowitz, S. J.; Margolis, Stephen E. (1994): Network Externality: An Uncommon Tragedy, in: *Journal of Economic Perspectives*, 8:2, 133-150.

Linn, Joshua (2006): *Stock Prices and the Cost of Environmental Regulation*, Working Paper: University of Illinois, Chicago.

Lintner, John (1965): Security Prices, Risk, and Maximal Gains From Diversification, in: *Journal of Finance*, 20:4, 587-615.

Liston, Catherine (1993): Price-cap versus rate-of-return regulation, in: *Journal of Regulatory Economics*, 5:1, 25-48.

Lohmann, Larry (2009): Toward a different debate in environmental accounting: The cases of carbon and cost-benefit, in: *Accounting, Organizations and Society*, 34:3/4, 499-534.

Lohmann, Martin (1964): *Einführung in die Betriebswirtschaftslehre*, 4th ed., Tübingen: Mohr (Siebeck).

London Economics (2007): *Structure and Performance of Six European Wholesale Electricity Markets in 2003, 2004 and 2005*, Report to the DG Comp.

Lorson, Peter; Töbe, Marc (2008): Bilanzierungsfeld Emissionsrechtehandel, in: *Zeitschrift für internationale und kapitalmarktorientierte Rechnungslegung*, 8:7-8, 498-510.

Lücke, Wolfgang (1955): Investitionsrechnung auf der Basis von Ausgaben oder Kosten?, in: *Zeitschrift für handelswirtschaftliche Forschung*, 7:7, 310-324.

Luft, Joan; Shields, Michael D. (2002): *Mapping Management Accounting: Graphics and Guidelines for Theory-Consistent Empirical Research*, Working Paper: Michigan State University.

Luhmann, Hans-Jochen (2008): Das statuierte Exempel: Die EU reguliert internationale Luftfahrt hinsichtlich ihres Klimaeffekts unilateral, in: *Zeitschrift für Energiewirtschaft*, 32:4, 230-234.

Luthardt, Ulf; Zimmermann, Jochen (2009): A European view on the legitimacy of accounting procedures: Towards a deliberative-accountability framework for analysis, in: *Research in Accounting Regulation*, 21:2, 79-88.

Lyon, Randolph M. (1982): Auctions and Alternative Procedures for Allocating Pollution Rights, in: *Land Economics*, 58:1, 16-32.

MacKenzie, Donald (2009): Making things the same: Gases, emission rights and the politics of carbon markets, in: *Accounting, Organizations and Society*, 34:3-4, 440-455.

Makhija, Mona (2003): Comparing the Resource-Based and the Market-Based Views of the Firm: Empirical Evidence form Czech Privatization, in: *Strategic Management Journal*, 24:5, 433-451.

Malueg, David A. (1989): Emission credit trading and the incentive to adopt new pollution abatement technology, in: *Journal of Environmental Economics and Management*, 16:1, 52-57.

Mandell, Svante (2005): The choice of multiple or single auctions in emission trading, in: *Climate Policy*, 5:1, 97-107.

Mansanet-Bataller, Maria; Pardo, Ángel; Valor, Enric (2007): CO_2 Prices, Energy and Weather, in: *Energy Journal*, 28:3, 73-92.

Marquardt, Carol A.; Wiedman, Christine I. (2004): The Effect of Earnings Management on the Value Relevance of Accounting Information, in: *Journal of Business Finance & Accounting*, 31:3/4, 297-332.

Marschak, Jacob (1971): Economics of Information Systems, in: *Journal of the American Statistical Association*, 66:333, 192-219.

Marshall, Ronald M. (1972): Determining an Optimal Accounting Information System for an Unidentified User, in: *Journal of Accounting Research*, 10:2, 286-307.

Mattessich, Richard (2006): The Information Economic Perspective of Accounting: Its Coming of Age, in: *Canadian Accounting Perspectives/ Perspectives Comptables Canadiennes*, 5:2, 209-236.

McWatters, Cheryl S.; Zimmerman, Jerold L.; Morse, Dale C. (2008): *Management Accounting: Analysis and Interpretation*, Harlow: Pearson Educational.

Meeus, Leonardo; Purchala, Konrad; Belmans, Ronnie (2005): Development of the Internal Electricity Market in Europe, in: *Electricity Journal*, 6:18, 25-35.

Mellerowicz, Konrad (1973): *Kosten und Kostenrechnung*, 5th ed., Berlin et al.: De Gruyter.

Menger, Carl (1968): *Grundsätze der Volkswirtschaftslehre*, 2nd ed., Tübingen: Mohr (Siebeck).

Menrad, Siegfried (1965): *Der Kostenbegriff*, Berlin: Duncker & Humblot, Habilitation, University of Tübingen.

Merchant, Kenneth A. (1990): The Effects of Financial Control on Data Manipulation and Management Myopia, in: *Accounting, Organizations and Society*, 15:4, 297-313.

Merton, Robert C. (1973): An Intertemporal Capital Asset Pricing Model, in: *Econometrica*, 41:5, 867-887.

Metzger, Axel (2008): Energiepreise auf dem Prüfstand: Zur Entgeltkontrolle nach Energie-, Kartell- und Vertragsrecht, in: *Zeitschrift für das gesamte Handelsrecht und Wirtschaftsrecht*, 172:4, 458-477.

Migué, Jean-Luc; Marceau, Richard (1993): Pollution Taxes, Subsidies, and Rent Seeking, in: *Canadian Journal of Economics/Revue canadienne d'Économique*, 26:2, 355-365.

Milliman, Scott R.; Prince, Raymond (1989): Firm incentives to promote technological change in pollution control, in: *Journal of Environmental Economics and Management*, 17:3, 247-265.

Milne, Markus J. (1996): Capitalizing and Appropriating Society's Rights to Clean Air: A Comment on Wambsganss & Sanford's Accounting Proposal, in: *Critical Perspectives on Accounting*, 7:6, 681-695.

Mishan, E. J. (1971): The Postwar Literature on Externalities: An Interpretative Essay, in: *Journal of Economic Literature*, 9:1, 1-28.

Mitchell, William C. (1988): Virginia, Rochester, and Bloomington: Twenty-five years of public choice and political science, in: *Public Choice*, 56:2, 101-119.

Mitchell, William C.; Munger, Michael C. (1991): Economic Models of Interest Groups: An Introductory Survey, in: *American Journal of Political Science*, 35:2, 512-546.

Montgomery, W. David (1972): Markets in Licenses and Efficient Pollution Control Programs, in: *Journal of Economic Theory*, 5:3, 395-418.

Mossin, Jan (1966): Equilibrium in a Capital Asset Market, in: *Econometrica*, 34:4, 768-783.

Müller-Christ, Georg; Hülsmann, Michael (2003): Quo vadis Umweltmanagement?: Entwicklungsperspektiven einer nachhaltigkeitsorientierten Managementlehre, in: *Die Betriebswirtschaft*, 63:3, 257-277.

Muller, Aline; Verschoor, Willem F. C. (2006): European Foreign Exchange Risk Exposure, in: *European Financial Management*, 12:2, 195-220.

Munzel, Gerhard (1966): *Die fixen Kosten in der Kostenträgerrechnung*, Wiesbaden: Gabler.

Nakano, Isao (1972): Noise and Redundancy in Accounting Communication, in: *Accounting Review*, 47:4, 693-708.

Neuhoff, Karsten; Åhman, Markus; Betz, Regina; Cludius, Johanna; Ferrario, Federico; Holmgren, Kristina; Pál, Gabriella; Grubb, Michael; Matthes, Felix; Rogge, Karoline; Sato, Misato; Schleich, Joachim; Sijm, Jos; Türk, Andreas; Kettner, Claudia; Walker, Neil (2006a): Implications of announced phase II national allocation plans for the EU ETS, in: *Climate Policy*, 6:4, 411-422.

Neuhoff, Karsten; Keats Martinez, Kim; Sato, Misato (2006b): Allocation, incentives and distortions: the impact of the EU ETS emissions allowance allocations to the electricity sector, in: *Climate Policy*, 6:1, 73-91.

Newbery, David M. (2001): Economic reform in Europe: integrating and liberalizing the market for services, in: *Utilities Policy*, 10:2, 85-97.

Nicklisch, Heinrich (1932): *Die Betriebswirtschaft*, 7th ed., Stuttgart: Poeschel.

Noble, H. S. (1926): Some Cost Accounting Concepts of Wider Application, in: *Accounting Review*, 1:4, 48-54.

Nordhaus, William D. (1973): The Allocation of Energy Resources, in: *Brookings Papers on Economic Activity*, 1973:3, 529-576.

Norton, Seth W. (1985): Regulation and Systematic Risk: The Case of Electric Utilities, in: *Journal of Law and Economics*, 28:3, 671-686.

Nydahl, Stefan (1999): Exchange rate exposure, foreign involvement and currency hedging of firms: some Swedish evidence, in: *European Financial Management*, 5:2, 241-257.

Nye, Michael; Owens, Susan (2008): Creating the UK Emission Trading Scheme: Motives and Symbolic Politics, in: *European Environment*, 18:1, 1-15.

Oberender, Peter; Rüter, Georg (1993): The steel industry: a crisis of adaption, in: de Jong, Hendrik Wouter (Ed.): *The Structure of European Industry*, 3rd ed., Dordrecht: Kluwer, 65-90.
Oberndorfer, Ulrich (2009): EU Emission Allowances and the stock market: Evidence from the electricity industry, in: *Ecological Economics*, 68:4, 1116-1126.
Ohlson, James A. (1995): Earnings, Book Values, and Dividends in Equity Valuation, in: *Contemporary Accounting Research*, 11:2, 661-687.
Olson, Mancur (1980): *The Logic of Collective Action: Public Goods and the Theory of the Group*, 8th ed., Cambridge (US): Harvard University Press.
Paolella, Marc S.; Taschini, Luca (2008): An econometric analysis of emission allowance prices, in: *Journal of Banking and Finance*, 32:10, 2022-2032.
Parsons, John E.; Ellerman, A. Denny; Feilhauer, Stephan (2009): Designing a U.S. Market for CO_2, in: *Journal of Applied Corporate Finance*, 21:1, 79-86.
Pedersen, Sigurd Lauge (2001): The Danish CO_2 Emissions Trading System, in: *Reciel*, 9:3, 223-231.
Peltzman, Sam (1976): Toward a More General Theory of Regulation, in: *Journal of Law and Economics*, 19:2, 211-240.
Pierson, Paul (2000): The Limits of Design: Explaining Institutional Origins and Change, in: *Governance: An International Journal of Policy, Administration and Institutions*, 13:4, 475-499.
Pigou, Arthur Cecil (1962): *The Economics of Welfare*, 4th ed. reprint, London: Macmillan.
Pitelis, Christos (1994): On the nature of the capitalist state, in: *Review of Political Economy*, 6:1, 72-105.
Pollack, Mark A. (1994): Creeping Competence: The Expanding Agenda of the European Community, in: *Journal of Public Policy*, 14:2, 95-145.
Porter, Michael E. (1991): Towards a Dynamic Theory of Strategy, in: *Strategic Management Journal*, 12:8, 95-117.
Posner, Richard A. (1974): Theories of Economic Regulation, in: *Bell Journal of Economics and Management Science*, 5:2, 335-358.
Preinreich, Gabriel A. D. (1938): Annual Survey of Economic Theory: The Theory of Depreciation, in: *Econometrica*, 6:3, 219-241.
Puxty, A. G.; Willmott, Hugh C.; Cooper, David J.; Lowe, Tony (1987): Modes of regulation in advanced capitalism: Locating accountancy in four countries, in: *Accounting, Organizations and Society*, 12:3, 273-291.
Quiggin, John (1988): Private and Common Property Rights in the Economics of the Environment, in: *Journal of Economic Issues*, 22:4, 1071-1087.
Quirion, Philippe (2005): Does uncertainty justify intensity emission caps?, in: *Resource and Energy Economics*, 27:4, 343-353.

Rahimi, Farrokh A.; Sheffrin, Anjali Y. (2003): Effective Market Monitoring in Deregulated Electricity Markets, in: *IEEE Transactions on Power Systems*, 18:2, 486-493.

Rappaport, Alfred (1981): Selcting strategies that create shareholder value, in: *Harvard Business Review*, 59:3, 139-149.

Reichelstein, Stefan (1997): Investment Decisions and Managerial Performance Evaluation, in: *Review of Accounting Studies*, 2:2, 157-180.

Requate, Till (2005): Dynamic incentives by environmental policy instruments – a survey, in: *Ecological Economics*, 54:2-3, 175-195.

Riddick, Leigh A. (1992): The effects of regulation on stochastic systematic risk, in: *Journal of Regulatory Economics*, 4:2, 139-157.

Riebel, Paul (1994): Core features of the "Einzelkosten- und Deckungsbeitragsrechnung", in: *European Accounting Review*, 3:3, 515-546.

Rogge, Karoline; Schleich, Joachim; Betz, Regina (2006): *An Early Assessment of National Allocation Plans for Phase 2 of EU Emission Trading*, Working Paper: Fraunhofer Institute Systems and Innovation Research, Karlsruhe, and University of New South Wales, Sydney.

Rogler, Silvia (2005): Bilanzierung von CO_2-Emissionsrechten, in: *Zeitschrift für internationale und kapitalmarktorientierte Rechnungslegung*, 5:6, 255-263.

Rogler, Silvia; Lange, Diana; Straub, Sandro Veit (2009): Bilanzierung von Emissionsrechten: Eine Fallstudie zu Ansatz- und Bewertungsfragen nach IFRS, in: *Zeitschrift für internationale und kapitalmarktorientierte Rechnungslegung*, 9:6, 371-382.

Ronen, Joshua (2008): To Fair Value or Not to Fair Value: A Broader Perspective, in: *Abacus*, 44:2, 181-208.

Roques, Fabien A.; Newbery, David M.; Nuttall, William J. (2008): Fuel mix diversification incentives in liberalized electricity markets: A Mean-Variance Portfolio theory approach, in: *Energy Economics*, 30:4, 1831-1849.

Ross, Stephen A. (1976): The Arbitrage Theory of Capital Asset Pricing, in: *Journal of Economic Theory*, 13:3, 341-360.

Roth, Ursula (1992): *Umweltkostenrechnung: Grundlagen und Konzeption aus betriebswirtschaftlicher Sicht*, Wiesbaden: Deutscher Universitäts-Verlag, Dissertation, University of Cologne.

Ryan, Stephen G.; Zarowin, Paul A. (2003): Why Has the Contemporaneous Linear Returns-Earnings Relation Declined?, in: *Accounting Review*, 78:2, 523-553.

Sadorsky, Perry (2001): Risk factors in stock returns of Canadian oil and gas companies, in: *Energy Economics*, 23:1, 17-28.

Samuelson, Paul A. (1937): A Note on Measurement of Utility, in: *Review of Economic Studies*, 4:2, 155-161.

Samuelson, Richard A. (1980): Should Replacement-Cost Changes Be Included in Income?, in: *Accounting Review*, 55:2, 254-268.

Samuelson, Richard A. (1996): The Concept of Assets in Accounting Theory, in: *Accounting Horizons*, 10:3, 147-157.

Sandmo, Agnar (1971): On the Theory of Competitive Firm Under Price Uncertainty, in: *American Economic Review*, 61:1, 65-73.

Sawkins, John W. (1996): Balancing Multiple Interests in Regulation: An Event Study of the English and Welsh Water Industry, in: *Journal of Regulatory Economics*, 9:3, 249-268.

Schaltegger, Stefan; Sturm, Andreas (1992): *Ökologieorientierte Entscheidungen in Unternehmen: Ökologisches Rechnungswesen statt Ökobilanzierung: Notwendigkeit, Kriterien, Konzepte*, Bern: Paul Haupt, Dissertation, University of Basel.

Schmalenbach, Eugen (1956): *Kostenrechnung und Preispolitik*, 7th ed., Cologne et al.: Westdeutscher Verlag.

Schmalenbach, Eugen (1962): *Dynamische Bilanz*, 13th ed., Cologne et al.: Westdeutscher Verlag.

Schmidt, Reinhard H.; Terberger, Eva (2003): *Grundzüge der Investitions- und Finanzierungstheorie*, 4th ed. (reprint), Wiesbaden: Gabler.

Schneider, Dieter (1961): Kostentheorie und verursachungsgemäße Kostenrechnung, in: *Zeitschrift für handelswissenschaftliche Forschung*, 13:10/11, 677-707.

Schneider, Dieter (1994): *Betriebswirtschaftslehre – Band 2: Rechnungswesen*, Munich et al.: Oldenbourg.

Schneider, Uwe A.; McCarl, Bruce A. (2003): Economic Potential of Biomass Based Fuels for Greenhouse Gas Emission Mitigation, in: *Environmental and Resource Economics*, 24:4, 291-312.

Scholz, John T. (1991): Cooperative Regulatory Enforcement and the Politics of Administrative Effectiveness, in: *American Political Science Review*, 85:1, 115-136.

Schreiner, Manfred (1990): Ökologische Herausforderungen an die Kosten- und Leistungsrechnung, in: Freimann, Jürgen (Ed.): *Ökologische Herausforderungen an die Betriebswirtschaftslehre*, Wiesbaden: Gabler, 197-214.

Schuppert, Gunnar Folke (1990): Grenzen und Alternativen von Steuerung durch Recht, in: Grimm, Dieter (Ed.): *Wachsende Staatsaufgaben – sinkende Steuerungsfähigkeit des Rechts*, Baden-Baden: Nomos, 217-249.

Schweitzer, Marcell; Küpper, Hans-Ulrich (1989): *Systeme der Kosten- und Erlösrechnung*, 7th ed., Munich: Vahlen.

Scitovsky, Tibor (1954): Two Concepts of External Economies, in: *Journal of Political Economy*, 62:2, 143-151.

Serrallés, Roberto J. (2006): Electric energy restructuring in the European Union: Integration, subsidiarity and the challenge of harmonization, in: *Energy Policy*, 34:16, 2542-2551.

Shannon, Claude Elwood; Weaver, Warren (1998): *The Mathematical Theory of Communication*, Urbana et al.: University of Illinois Press.

Sharpe, William F. (1963): A Simplified Model for Portfolio Analysis, in: *Management Science*, 9:2, 277-293.

Shirley, Mary M. (1999): Bureaucrats in business: The roles of privatization versus corporatization in state-owned enterprise reform, in: *World Development*, 27:1, 115-136.

Siebenhüner, Bernd (2003): The changing role of nation states in international environmental assessments – the case of the IPCC, in: *Global Environmental Change*, 13:2, 113-123.

Sijm, Jos; Neuhoff, Karsten; Chen, Yihsu (2006): CO_2 cost pass-through and windfall profits in the power sector, in: *Climate Policy*, 6:1, 49-72.

Simon, Herman Veit (1899): *Die Bilanzen der Aktiengesellschaften und der Kommanditgesellschaften auf Aktien*, 3rd ed., Berlin: Guttentag.

Simpson, R. David (1995): Optimal Pollution Taxation in a Cournot Duopoly, in: *Environmental and Resource Economics*, 6:4, 359-369.

Sioshansi, Fereidoon P. (2006): Electricity market reform: What has the experience taught us thus far?, in: *Utilities Policy*, 14:2, 63-75.

Skjærseth, Jon; Wettestad, Jørgen (2008): Implementing EU emissions trading: success or failure?, in: *International Environmental Agreements: Politics, Law and Economics*, 8:3, 275-290.

Smale, Robert; Hartley, Murray; Hepburn, Cameron; Ward, John; Grubb, Michael (2006): The impact of CO_2 emissions trading on firm profits and market prices, in: *Climate Policy*, 6:1, 31-48.

Smith, Bryant Walker (2009): *Stakeholder Reaction to Emissions Trading in the United States, the European Union, and the Netherlands*, Working Paper: New York University School of Law.

Smith, Clifford W.; Watts, Ross L. (1992): The investment opportunity set and corporate financing, dividend, and compensation policies, in: *Journal of Financial Economics*, 32:3, 263-292.

Snyder, James M. (1991): On Buying Legislatures, in: *Economics & Politics*, 3:2, 93-109.

Söderholm, Patrik (2001): Fossil fuel flexibility in west European power generation and the impact of system load factors, in: *Energy Economics*, 23:1, 77-97.

Spiller, Pablo T. (1990): Politicians, Interest Groups, and Regulators: A Multiple-Principals Agency Theory of Regulation, or "Let Them Be Bribed", in: *Journal of Law and Economics*, 33:1, 65-101.

Stalon, Charles G. (1997): Electric industry governance: Reconciling competitive power markets and the physics of complex transmission interconnections, in: *Resource and Energy Economics*, 19:1-2, 47-83.

Stavins, Robert N. (1998): What Can We Learn from the Grand Policy Experiment? Lessons from SO_2 Allowance Trading, in: *Journal of Economic Perspectives*, 12:3, 69-88.

Stigler, George J. (1971): The Theory of Economic Regulation, in: *Bell Journal of Economics and Management Science*, 2:1, 3-21.

Streeck, Wolfgang; Schmitter, Philippe C. (1985): Community, market, state – and associations?: The prospective contribution of interest governance to social order, in: *European Sociological Review*, 1:2, 119-138.

Suh, Yoon S. (1990): Communication and Income Smoothing Through Accounting Method Choice, in: *Management Science*, 36:6, 704-723.

Sun, Jeanne-Mey; Pelkmans, Jacques (1995): Regulatory Competition in the Single Market, in: *Journal of Common Market Studies*, 33:1, 67-89.

Sunder, Shyam (1973): Relationship between Accounting Changes and Stock Prices: Problems of Measurement and Some Empirical Evidence, in: *Journal of Accounting Research*, 11 (Empirical Research in Accounting: Selected Studies 1973), 1-45.

Teulings, Coen; Hartog, Joop (1998): *Corporatism or competition? Labour contracts, institutions and wage structures in international comparison*, Cambridge (UK): Cambridge University Press.

Thomas, Steve (2003): The Seven Brothers, in: *Energy Policy*, 31:5, 393-403.

Thurm, Ralph A. (1997): Kostentransparenz für das betriebliche Umweltmanagement, in: *Kostenrechnungspraxis*, 41:6, 328-334.

Tietenberg, Thomas H. (1990): Economic Instruments for Environmental Regulation, in: *Oxford Review of Economic Policy*, 6:1, 17-33.

Tietenberg, Thomas H. (2003): The Tradable Permits Approach to Protecting the Commons: Lessons for Climate Change, in: *Oxford Review of Economic Policy*, 19:3, 400-419.

Tirole, Jean (1986): Hierarchies and Bureaucracies: On the Role of Collusion in Organizations, in: *Journal of Law, Economics & Organization*, 2:2, 181-214.

Trück, Stefan; Borak, Szymon; Härdle, Wolfgang; Weron, Rafal (2006): *Convenience Yields for CO_2 Emission Allowance Futures Contracts*, Working Paper: Queensland University of Technology, Brisbane, et al.

Truman, David B. (1967): *Governmental Process: Political Interests and Public Opinion*, 14th ed., New York: Alfred A. Knopf.

Tufano, Peter (1998): The Determinants of Stock Price Exposure: Financial Engineering and the Gold Mining Industry, in: *Journal of Finance*, 53:3, 1015-1052.

van Koten, S.; Ortmann, A. (2008): The unbundling regime for electricity utilities in the EU: A case of legislative and regulatory capture?, in: *Energy Economics*, 30:6, 3128-3140.

van Vuuren, D. P.; Cofala, J.; Eerens, H. E.; Oostenrijk, R.; Heyes, C.; Klimont, Z.; den Elzen, M. G. J.; Amann, M. (2006): Exploring the ancillary benefits of the Kyoto Protocol for air pollution in Europe, in: *Energy Policy*, 34:4, 444-460.

van Zeben, Josefien A. W. (2009): *Law-making in the Revised EU ETS – A Study in Regulatory Federalism*, Working Paper: Universiteit van Amsterdam.

Veith, Stefan; Werner, Jörg R.; Zimmermann, Jochen (2009): Capital market response to emission rights returns: Evidence from the European power sector, in: *Energy Economics*, 31:4, 605-613.

Veith, Stefan; Werner, Jörg R.; Zimmermann, Jochen (2010): *Competing Accounting Treatments for Emission Rights: A Capital Market Perspective*, Working Paper: University of Bremen.

Verbruggen, Aviel (2008): Windfalls and other profits, in: *Energy Policy*, 36:9, 3249-3251.

Verde, Stefano (2008): Everybody merges with somebody – The wave of M&As in the energy industry and the EU merger policy, in: *Energy Policy*, 36:3, 1125-1133.

Vernimmen, Pierre (2005): *Corporate Finance: Theory and Practice*, Chichester et al.: Wiley.

Viner, Jacob (1932): Cost Curves and Supply Curves, in: *Zeitschrift für Nationalökonomie*, 1:3, 23-46.

Völker-Lehmkuhl, Katharina (2006): *Praxis der Bilanzierung und Besteuerung von CO_2-Emissionsrechten*, Berlin: Erich Schmidt Verlag.

Völker-Lehmkuhl, Katharina; Lösler, Andreas (2005): CO_2-Emissionsrechte – Bilanzielle Behandlung nach US-GAAP, in: *Der Betrieb*, 58:9, 457-461.

Volmer, Philipp B.; Werner, Jörg R.; Zimmermann, Jochen (2007): New governance modes for Germany's financial reporting system: another retreat of the nation state?, in: *Socio-Economic Review*, 5:3, 437-465.

von der Fehr, Nils-Henrik; Amundsen, Eirik S.; Bergman, Lars (2005): *The Nordic Market: Signs of Stress?*, Working Paper: University of Cambridge.

von Wieser, Friedrich (1884): *Über den Ursprung und die Hauptgesetze des wirthschaftlichen Werthes*, Vienna: Alfred Hölder.

Wagenhofer, Alfred; Ewert, Ralf (2007): *Externe Unternehmensrechnung*, 2nd ed., Berlin et al.: Springer.

Wagner, Karin; Vassilopoulos, Michalis (2000): *The European Cement Industry*, Background Assessment for the IPTS BAT-Competitiveness Project.

Walker, Jack L. (1983): The Origins and Maintenance of Interest Groups in America, in: *American Political Science Review*, 77:2, 390-406.

Wambsganss, Jacob R.; Sanford, Brent (1996): The Problem with Reporting Pollution Allowances, in: *Critical Perspectives on Accounting*, 7:6, 643-652.

Watts, Ross L. (2003): Conservatism in Accounting – Part I: Explanations and Implications, in: *Accounting Horizons*, 17:3, 207-221.

Watts, Ross L.; Zimmerman, Jerold L. (1978): Towards a Positive Theory of the Determination of Accounting Standards, in: *Accounting Review*, 53:1, 112-134.

Watts, Ross L.; Zimmerman, Jerold L. (1979): The Demand for and Supply of Accounting Theories: The Market for Excuses, in: *Accounting Review*, 54:2, 273-305.

Weber, Matthias; Hoogma, Remco (1998): Beyond National and Technological Styles of Innovation Diffusion: A Dynamic Perspective on Cases from the Energy and Transport Sectors, in: *Technology Analysis & Strategic Management*, 10:4, 545-566.

Weersink, Alfons; Livernois, John; Jason, F. Shogren; Shortle, James S. (1998): Economic Instruments and Environmental Policy in Agriculture, in: *Canadian Public Policy/Analyse de Politiques*, 24:3, 309-327.

Weiss, Leonard W. (1979): The Structure-Conduct-Performance Paradigm and Antitrust, in: *University of Pennsylvania Law Review*, 127:4, 1104-1140.

White, Halbert (1980): A Heteroskedasticity-Consistent Covariance Matrix Estimator and a Direct Test for Heteroskedasticity, in: *Econometrica*, 48:4, 817-838.

Woerdman, Edwin (2000): Implementing the Kyoto protocol: why JI and CDM show more promise than international emissions trading, in: *Energy Policy*, 28:1, 29-38.

Woo, Chi-Keung; Lloyd, Debra; Tishler, Asher (2003): Electricity market reform failures: UK, Norway, Alberta and California, in: *Energy Policy*, 31:1103-1115.

Wurzel, Rüdiger K. W.; Brückner, Lars; Jordan, Andrew; Zito, Anthony (2003): Struggling to Leave Behind a Highly Regulatory Past? 'New' Environmental Policy Instruments in Austria, in: *Environmental Politics*, 12:1, 51-72.

Zeff, Stephen A. (1962): Replacement Cost: Member of the Family, Welcome Guest or Intruder?, in: *Accounting Review*, 37:4, 611-625.

Zeff, Stephen A. (1978): The Rise of "Economic Consequences", in: *Journal of Accountancy*, 146:6, 56-63.

Ziegler, Andreas; Schröder, Michael; Schulz, Anja; Stehle, Richard (2007): Multifaktormodelle zur Erklärung deutscher Aktienrenditen: Eine empirische Analyse, in: *Zeitschrift für betriebswirtschaftliche Forschung*, 59:3, 355-389.

Zimmerman, Jerold L. (2001): Conjectures regarding empirical managerial accounting research, in: *Journal of Accounting and Economics*, 32:1-3, 411–427.

Zimmermann, Jochen (2006): Die Bilanzierung von Emissionsrechten in Informations-, Handels- und Steuerbilanzen – Ein kritischer Vergleich, in: *Steuern und Bilanzen*, 8:10, 367-375.

Zimmermann, Jochen (2008): The Role of the Nation State in the Internationalization of Accounting Regimes, in: Hurrelmann, Achim; Leibfried, Stephan; Martens, Kerstin; Mayer, Peter (Eds.): *Transforming the Golden-Age Nation State*, Basingstoke: Palgrave Macmillan, 173-192.

Zimmermann, Jochen; Veith, Stefan (2007): Die Erfassung von CO_2-Emissionsrechten in der wertmäßigen und ökologischen Kostenrechnung, in: *Zeitschrift für Controlling und Management*, 51:5, 355-364.

Zimmermann, Jochen; Werner, Jörg R.; Volmer, Philipp B. (2008): *Global Governance in Accounting: Rebalancing Public Power and Private Commitment*, Basingstoke: Palgrave Macmillan.

Zúñiga, Gloria L. (1999): An Ontology of Economic Objects: An Application of Carl Menger's Ideas, in: *American Journal of Economics and Sociology*, 58:2, 299-312.

Peter Lang · Internationaler Verlag der Wissenschaften

Ignacio Pérez Domínguez

Greenhouse Gases: Inventories, Abatement Costs and Markets for Emission Permits in European Agriculture

A Modelling Approach

Frankfurt am Main, Berlin, Bern, Bruxelles, New York, Oxford, Wien, 2006.
213 pp., num. fig. and tab.
European University Studies: Series 5, Economics and Management. Vol. 3184
ISBN 978-3-631-55082-3 · pb. € 43,20*

The international community is putting some pressure on scientific researchers to come up with reliable indicators to estimate emissions from greenhouse gases. The development towards integrative multi gas strategy approaches has allowed the inclusion of the agricultural sector in the political agenda. Modelling alternatives for the estimation of emission factors, definition of policy instruments for greenhouse gas emission abatement as well as measurement of their economic effects are at this stage quite important for the coming multilateral negotiations.

Contents: Climate change and agriculture · Modelling of greenhouse gas emission inventories for the European agricultural sector · The use of emission abatement instruments: tradable permits as a viable option in agriculture · Calculation of marginal abatement cost curves · Modelling a market of tradable emission permits · Comparison between emission abatement instruments including price effects

Frankfurt am Main · Berlin · Bern · Bruxelles · New York · Oxford · Wien
Distribution: Verlag Peter Lang AG
Moosstr. 1, CH-2542 Pieterlen
Telefax 0041 (0) 32/376 17 27

*The €-price includes German tax rate
Prices are subject to change without notice

Homepage http://www.peterlang.de